PRAISE FOR
COYOTE AMERICA

*Finalist for the PEN/E. O. Wilson
Literary Science Writing Award*

Winner of the Sigurd Olson Nature Writing Award

Mountains and Plains Indie Bestseller

"Captivating. . . . Dan Flores looks at a creature whose howl sent shivers down the spines of generations of farmers and ranchers. They responded by waging war on an animal that not only refused to disappear, but began showing up in places like Central Park. The coyote turns out to be the Road Runner in disguise, and is having the last laugh after all."
—*Wall Street Journal*

"It is often impossible to separate how animals behave 'wild' from how they behave around humans. Coyotes are a startling example. . . . Historian Dan Flores has fun describing how coyotes make a mockery of our attempts to put nature in order: 'It turns out, the coyote really is The Dude, and The Dude absolutely abides.'"
—*New Scientist*

"Historian Dan Flores tracks the pedigree, chronicles the plight, and sings the praises of *Canis latrans* in his new book, *Coyote America: A Natural and Supernatural History*. Although his academic research is wide-ranging and his presentation nuanced, there's no doubt Flores's heart is on the side of the animal."
—*Daily Beast*

"A beautifully readable and meticulously researched book."
—Forbes.com

"A wonderful read . . . chock-full of detailed information and stories about this most adaptable mammal."
—*Psychology Today*

"Historian Flores has written about the American West for decades, so it's no surprise his gaze should turn to the region's scrappy mascot. Over the past 500 years, the original desert dweller has expanded its territory as far north as Alaska, south into the tropics, and deep into many cities. That ubiquity has created a host of problems for both the animal and its neighbors, human and otherwise. Flores captures all sides of the situation in this detailed portrait of an American icon."
—*Discover*

"A fascinating new biography of the species."—*Wild Life*

"Coyotes have a legendary appeal in North America, from the folklore tales of indigenous tribes to everyone's favorite 'Super Genius,' Wile E. Coyote. In *Coyote America*, Flores does more than just shed light on the legend; he explores 5 million years of biological history that lead up to the evolution of the modern coyote (*Canis latrans*) and details the unique versatility of an animal that has continued to thrive despite human campaigns of annihilation."
—*Chicago Tribune*

"The coyote should have been *TIME Magazine*'s Person of the Year. This deeply engrossing study is part scientific, part mythological, and part personal observation. It is fully fascinating."
—*Literary Hub*

"Wide-ranging, engaging, informative. . . . Flores is both a fine scholar and a most engaging writer. He argues most persuasively that we need to learn to live with coyote and the other beings with which we share this earth."
—*National Parks Traveler*

"Engaging. . . . Recommended."—*CHOICE*

"Engaging. . . . [Flores] provides a unique insight into the age-old war of man vs. wild. . . . [His] storytelling is riddled with humor and tidbits of information that pique interest and make it impossible to put down. . . . From cover to cover it's a truly thrilling read, leaving readers with a better idea of the way urban development has impacted the world and a desire to protect the existence of animals that have, in more recent days, been labeled dangerous predators. Acting as both a knowledgeable guide and a public-service announcement, this book is a must-read for anyone with an interest in animal welfare and environmentalism."—*Deseret News*

"A must-read book if you are interested in knowing more about this persecuted critter, revered by Native Americans long before the settlers arrived."—*Virginian Pilot*

"*Coyote America* possesses an extraordinary sweep and is an intriguing read."—*Albuquerque Journal*

"The most compelling part of Flores's story—and what makes *Coyote America* an important book—is what it says about the U.S. government's misguided and continuing battle against the coyote."
—*Santa Fe New Mexican*

"Flores stares long and deeply into the coyote's eyes, returning to us with cultural treasures both sparkling and lyrical."
—*Open Letters Monthly*

"A must-read for all Americans, whether you are a farmer or rancher, a suburban or city folk."—*Mother Earth News*

"A fascinating scientific and cultural history. . . . Deft prose and wide-ranging research do their part to carry Flores through the grimmer chapters of his narrative. . . . Whatever the coyote may still be wanting, that list no longer includes a book to do it justice."
—*New Mexico Magazine*

"Fascinating . . . essential literature in university courses on environmental studies, wildlife management, and general ecology and public policy. This book will appeal to ecologists as well as to a general audience seeking to better understand how modern humans have treated coyotes and build a new paradigm for a reformed and more holistic vision of how to manage coyotes with respect and compassion. . . . A copy of *Coyote America* should be given to all legislators to help in making informed and more cost-efficient and humane wildlife policies."—*Ecology*

"In a straightforward style, the author unpacks the myths and urban legends surrounding the coyote and conveys his admiration and respect for this incredibly intelligent predator. . . . Highly recommended for natural history enthusiasts interested in moving beyond the conventional wisdom about coyotes to gain a deeper understanding of their presence in our midst."—*Library Journal*

"A spirited blend of history, anthropology, folklore, and biology that is capable of surprises. . . . Well-written throughout and just the right length, Flores's book makes a welcome primer for living in a land in which coyotes roam freely—in, that is to say, the *Coyote America* of his title."—*Kirkus Reviews*

"Flores's mix of edification and entertainment is a welcome antidote to a creature so often viewed with fear."—*Publishers Weekly*

"In this brilliant book, Flores traces the wane and wax of the coyote. Their story is interwoven with our story, but it is also like our story, that of a species that has faced challenges and overcome them. Read this book if you want to understand the wild canids among us and also, perhaps, a little bit more about yourself."
—Rob Dunn, author of *The Man Who Touched His Own Heart*

"As I was reading *Coyote America* by Dan Flores, a coyote walked through our backyard. Magic occurs in these pages."
—Terry Tempest Williams, author of *The Hour of Land: A Personal Topography of America's National Parks*

"Dan Flores's *Coyote America* is an utterly fascinating look at the life and range of *Canis latrans*. It brilliantly blends environmental history with old-fashioned storytelling. Flores is a master of the American West and a personal hero. A must-read!"
—Douglas Brinkley, professor of history, Rice University, and author of *Rightful Heritage: Franklin D. Roosevelt and the Land of America*

"Think of *Coyote America* as a biography of our continent's most enigmatic and successful predator, but don't stop there. It is also a meditation, eloquent and insightful, on our relationship to wildlife, to nature, and even to our national culture. When you've read it, you won't sing the book's praises, you'll howl them."
—William deBuys, author of *The Last Unicorn* and *A Great Aridness*

"A biologist once told me, 'When the last man dies, a coyote will be howling over his grave.' This splendid book makes it clear why that's true, and why the persistent, enduring wildness of this remarkable neighbor should give us great delight."
—Bill McKibben, author of *Wandering Home*

"With a deft blend of science and history, Dan Flores shows us the coyote as trickster, survivor, and, ultimately, a reflection of ourselves. *Coyote America* paints a vivid and long-overdue portrait of an iconic animal. It's a terrific book."
—Thor Hanson, author of *The Triumph of Seeds* and *Feathers*

"A wily writer meets his natural subject. With erudition, pathos, and seductive humor, Dan Flores tells coyote stories that expose the animalism of Americans, and humans everywhere. The pleasure of his book is the cross-species love of being alive."
—Jared Farmer, author of *Trees in Paradise: A California History*

COYOTE AMERICA

ALSO BY DAN FLORES

American Serengeti: The Last Big Animals of the Great Plains

*Visions of the Big Sky: Painting and
Photographing the Northern Rocky Mountain West*

*The Natural West: Environmental History
in the Great Plains and Rocky Mountains*

*Horizontal Yellow:
Nature and History in the Near Southwest*

*Caprock Canyonlands:
Journeys into the Heart of the Southern Plains*

Canyon Visions: Photographs and Pastels of the Texas Plains

*Journal of an Indian Trader:
Anthony Glass and the Texas Trading Frontier*

*Southern Counterpart to Lewis and Clark:
The Freeman and Custis Expedition of 1806*

The Mississippi Kite: Portrait of a Southern Hawk

COYOTE AMERICA

A NATURAL AND SUPERNATURAL HISTORY

DAN FLORES

BASIC BOOKS
New York

Copyright © 2016 by Dan Flores

Hachette Book Group supports the right to free expression and the value of copyright. The purpose of copyright is to encourage writers and artists to produce the creative works that enrich our culture.

The scanning, uploading, and distribution of this book without permission is a theft of the author's intellectual property. If you would like permission to use material from the book (other than for review purposes), please contact permissions@hbgusa.com. Thank you for your support of the author's rights.

Basic Books
Hachette Book Group
1290 Avenue of the Americas, New York, NY 10104
www.basicbooks.com

Printed in the United States of America
First Trade Paperback Edition: September 2017

Published by Basic Books, an imprint of Perseus Books, LLC, a subsidiary of Hachette Book Group, Inc.

The Hachette Speakers Bureau provides a wide range of authors for speaking events. To find out more, go to www.hachettespeakersbureau.com or call (866) 376-6591.

The publisher is not responsible for websites (or their content) that are not owned by the publisher.

Print book interior design by Trish Wilkinson.

Library of Congress Cataloging-in-Publication Data
Names: Flores, Dan L. (Dan Louie), 1948–, author.
Title: Coyote America : a natural and supernatural history / Dan Flores.
Description: New York : Basic Books, 2016. | Includes index.
Identifiers: LCCN 2015043370 (print) | LCCN 2016001156 (ebook) | ISBN 9780465052998 (hardcover) | ISBN 9780465098538 (e-book)
Subjects: LCSH: Coyote—North America—History.
Classification: LCC QL737.C22 F63 2016 (print) | LCC QL737.C22 (ebook) | DDC 599.77/25—dc23
LC record available at http://lccn.loc.gov/2015043370

ISBN 978-0-465-09372-4 (paperback)

LSC-C

Printing 10, 2023

For Sara

Contents

INTRODUCTION	American Avatar	1
CHAPTER 1	Old Man America	21
CHAPTER 2	Prairie Wolf	53
CHAPTER 3	A War on Wild Things	81
CHAPTER 4	The Archpredator of Our Time	113
CHAPTER 5	Morning in America	151
CHAPTER 6	Bright Lights, Big Cities	189
CHAPTER 7	Coyote America	209
EPILOGUE	Coyote Consciousness	233
	Selected Bibliography	*249*
	Index	*257*

Few creatures on Earth possess a biography to match the coyote's.
Courtesy Dan Flores.

INTRODUCTION

American Avatar

Here is a vivid memory. I am on winter break from my university duties, and I am home visiting my parents in Louisiana. We are up early, stirring around in the kitchen in a gray, misty dawn in our hometown near the Black Bayou in the north of the state. My dad knows the paper will already be in the mailbox and asks me to walk out to the end of the driveway to fetch it in. I am in the act of pulling the *Shreveport Times* out of the box in silvery-blue twilight when I hear a rhythmic clatter and the scratching of toenails on the asphalt street. Plastic-wrapped newspaper in hand, I turn toward the commotion and find myself staring into the face of a coyote, which shoots me a quick sidelong glance as it passes by mere feet away, rocking along in that characteristic lope with which coyotes go through the world, close enough to pepper my ankles with dislodged gravel as it digs in to swerve past me. I register hot yellow eyes and something in its mouth. Maybe it's the neighbor's cat, although in the dim light it looks more like a trio of tallboys dangling from plastic six-pack rings.

Open-mouthed, I follow its course down the street, when a second coyote appears, weaving through three cars parked on the road opposite my parents' home. Pink tongue lolling, this one fixes me with a look that has something of a been-good-to-know-you wave as it passes, and in another couple of seconds it is at the corner intersection at the end of the block. There it pauses for an instant, turns to look back at me, then is gone in a movement I barely discern, like smoke dispersed by a sudden swirl of wind. Coyote number two seems empty-handed, and I register the thought (and tell my dad back in the kitchen), Ah, those looked like pull-tab beers anyway, so no bottle opener needed.

I tell this story not because it is unusual but for exactly the opposite reason. In twenty-first-century America likely a hundred scenes something like this unfold in towns and cities from coast to coast every morning. Close encounters with coyotes have now become the country's most common large-wildlife experience. They happen everywhere, in small towns like the one where I was raised and in gigantic cities: Los Angeles, San Francisco, Denver, Atlanta, Chicago, and now even New York City, the coyote's final frontier. A study of print media from 1998 to 2010 found a whopping 1,214 newspaper and magazine articles dedicated to human-coyote encounters in the United States for those years, and of course those included only the newsworthy events. Many thousands more like mine in Louisiana occurred without anyone calling a reporter. According to Stan Gehrt, Chicago's urban coyote guru ecologist, every major city in America now has a coyote-management study underway.

Our sense—and this is the way most media and online stories about coyotes have presented the matter—is that this is a brand-new phenomenon and something for which we humans are wholly responsible. Those sentiments reflect urban legend and human hubris. Urban coyotes are in no sense new in American history. And in truth, we are far less responsible than they are for their arrival everywhere among us now. In one of the myriad ways humans and coyotes eerily mimic one another, like us coyotes are a cosmopolitan species, able to live in a remarkable range of habitats. In our case, we have cleverly enhanced our own cosmopolitanism by taking our evolutionary habitat with us around the world: every North American home from Fairbanks to Tucson is heated or cooled to seventy-two degrees, the ambient temperature of the African setting where we evolved. Coyotes haven't bothered with recreating their desert and plains habitats wherever they've gone, but they have sought out prairies, habitat edges, and natural areas as they have colonized the continent, and they have assuredly used cleverness in spreading from their original home in the West across North America.

Coyotes have been "going along" (as the many Indian stories about them have always put it) for far longer than just the last century. Contemporary media accounts commonly assume that coyotes in eastern settings are an inexplicable phenomenon of the past few decades, that they are "invaders" of the country east of the Mississippi, the result of a sort of reverse-direction coyote manifest destiny. But that relentless coyote trot and sidelong glance has been eating up global space for a very long time. The ancestral canids that would eventually produce coyotes sprang from North American stock, a line of animals that evolved in the American Southwest. That ancient coyote line spawned animals that migrated to Eurasia and eventually to Africa to become Old World jackals. In North America, archeological sites from the late Pleistocene have yielded coyote remains from as far east as Pennsylvania and West Virginia, and genetic evidence indicates that coyotes thronged eastward out of their core range in the American West in at least two swarms, roughly between three hundred and nine hundred years ago. The truth is, roaming coyotes have probably been swimming the Mississippi River to eastern America during most decades since there have been coyotes.

Five hundred years ago, when Old Worlders first came to America, coyotes largely hewed to a range centered on the deserts and plains of the interior West. The initial European descriptions of coyotes fixed their range in the north as the prairies of the modern Canadian provinces of Manitoba, Saskatchewan, and Alberta. Farther west, fur traders found them in southern British Columbia, but not north of there, and they were nowhere along the Canadian coast. Southward down the Pacific shore, there were no coyotes in coastal Washington or Oregon, but as the country dried farther south, they appeared in coastal California all the way down to Baja.

In the interior West, coyotes were not numerous in mountain ranges like the Sierras, Cascades, or Rockies, but apparently a few did roam the high mountains. In all the open country of the interior, though, everywhere there were deserts or plains, as far eastward as Minnesota and the Blackland Prairies of Oklahoma and East Texas, coyotes lived in the millions five hundred years ago.

At some point in their history, coyotes pushed southward down the continent, spilling out of the Sonoran and Chihuahuan deserts all the way to central Mexico, and human activity may well have drawn them there. Thousands of years of large-scale landscape modification by agricultural societies in that part of Mesoamerica culminated 800 to 1,000 years ago in the Aztec Empire, whose sprawling cities seem to have effected a magnetic pull on coyotes, luring them down from their northern homelands. The lucrative possibilities that life among humans offered coyotes only multiplied when Europeans arrived with herds of clumsy, dim-witted domesticated sheep and goats from the Old World. In the late 1500s the little desert dogs initiated their first recorded range expansion following European arrival when they began to trot southward from the old Aztec capital down through Mexican states like Oaxaca, seeking easy prey among the pastoral flocks. From there they pressed on toward the equator, loping all the way into Guatemala, Honduras, and El Salvador. Spanish chroniclers reported the presence of a wolflike animal known to the Indians as *coyotl* as early as 1600.

Another European activity in wild North America, the fur trade, probably provoked coyotes' next famous range expansion. By the 1830s Canadian traders working westward from Hudson's Bay began to see coyotes north of where they'd ever noticed them before. Decade by decade, hard-bitten trappers in the frozen North wrote to their superiors that the small "prairie wolves" unaccountably seemed to be edging farther north, until by the end of the nineteenth century there were coyotes in the Northwest Territories and even 'round about Whitehorse in the Yukon and Anchorage in Alaska. That amounted to a range expansion of more than six hundred miles across the previous three-quarters of a century. Then, famously, the scavenging possibilities presented by dead horses and mules during Far North mining rushes led coyotes to follow along in the wake of miners plying the Klondike Trail. By 1930 there were coyotes north of the Yukon River, all the way to the Brooks Range. In both the Yukon and Alaska, a few even made it past the Arctic Circle, where Eskimo people, confounded by their presence, speculated among themselves that they were some kind of never-before-seen fox.

All this is to say that long before America's desert canines began to show up in Chicago and Manhattan, across the four centuries between the 1520s and the 1930s, coyotes had added about 2,000 miles of territory on the southern and northern edges of their range, which now spanned 7,500 miles of the continent and included every habitat from tropical jungle to frozen tundra. Yet the little western song-dogs were just getting started. By 1950 coyotes had begun to spill across the western mountains into the coastal country of the Pacific, showing up on the outskirts of Seattle and Portland, places where no one had ever seen a coyote before. At the same time, 1,500 miles to the east, they were colonizing almost all of Ontario, as far toward the Atlantic as Toronto and Ottawa. By the mid-twentieth century, the United States and Canada, from the Pacific to east of the Great Lakes, had become coyote country. To the 2,000 north-south miles of new coyote territory since the coming of Europeans, coyotes had now added nearly 1,000 additional miles west and east of where the British, French, and Spanish had first found them. Expansion by only one other large mammal matches this in American history. We humans did it twice: first, when the original Indian hunter-settlers poured out of Siberia all the way to the tip of South America in fewer than 1,000 years; second, during our so-called manifest destiny, a colonization that took Europeans from footholds along the Atlantic Seaboard and in the Southwest across virtually all of North America in about 300 years.

Across the history of life on Earth, animals and birds of many species have routinely colonized new country. That's enough of a marker of adaptive success that biologists apply the term "cosmopolitan" to species that are especially flexible regarding the habitats where they can live. Evolving in America, the ancestors of horses spread across Asia, Europe, and Africa, where they became zebras and quaggas. Bovine evolution in Southeast Asia eventually brought bison to North America, where they became an icon of the continent. But the range expansion of a wild animal for thousands of miles in every direction, often through dense settlements of humans who in recent history have been committed to that animal's eradication, is truly remarkable. A suite of factors must be involved.

The intelligence and flexibility that evolution bequeathed this small wolf was undoubtedly most important of all. Southwestern Hispanos have a rich folk tradition about coyotes and have long said that the only thing smarter than a coyote is God. It is a certainty that only humans and a handful of other species are capable of the variety of lifestyles coyotes can lead, from living with a pack and cooperating cleverly to attain group goals to slipping into the cracks of the world to fend for themselves as lone individuals. But other factors were at play. An obvious one from their history is that coyotes—at least some coyotes—not only survive among humans but have long quested after opportunities among us as a part of their evolving way of life. Unlike so many other wild animals, coyotes have been seeking humans out from the time we arrived in America, almost "testing" how closely they and we can function in the world. This striking trait has created an unusual history.

We also aided them in ways we little understood at the time. One of the "pull" factors in drawing coyotes out of the West seems to have been our almost complete extirpation of eastern and southern wolves. Since colonial times Americans had waged war against *Canis lycaon*, the wolf of the Northeast, and *Canis rufus*, the red wolf of the South and Atlantic Seaboard. By the early 1900s we had pushed these wolf species to the edge of extinction almost everywhere east of the Mississippi River. Western coyotes venturing into these landscapes, where all sign and scent of their ancient relatives was absent or fading, found whole new possibilities available to a midsize canid predator. East of the Mississippi they found wet, humid, forested landscapes drastically unlike their natal deserts and plains, with different prey and food sources. But by the twentieth century, eastern landscapes like those in New England were recovering from intensive farming and clearing, once again reforesting and rewilding. Adirondack Park in upstate New York, new national forests in New Hampshire, Vermont, and the Appalachians, and new national parks like Great Smoky Mountains gave the East some essential wildlands in the twentieth century. East of the big river were far more people, roads, towns, and cities. But a species that had survived the Pleistocene Extinctions, consorted with humans before, and escaped a scorched-earth campaign against them in the West evidently found none of that intimidating.

Because naturally there was also a "push" factor in coyotes' takeover of America. For a century before lone animals and pairs of coyotes began to show up in the cities of the Eastern Seaboard, coyotes had been the special targets and victims of a crusade against their kind that surpassed any other in recorded history in terms of the range of killing techniques and cruelty. It was a war of extermination, publicly proclaimed, and its weapons had progressed from guns and traps and dogs to saturation poisoning with a range of predacides invented specifically for the purpose. Wolves fell quickly before this onslaught, but, amazingly, decades of intensive persecution, culminating in a western landscape littered with millions of poison baits, did not eradicate coyotes. Yet the unrelenting pressure on them did invoke an ancient coyote biological imperative: it triggered larger litters of pups and colonization behavior that pushed them into new settings everywhere on the outer margins of their core range.

So starting in the 1920s, coyotes began suddenly and mysteriously showing up in places east of the Mississippi River where Americans had never seen them before. With coyotes rapidly colonizing Ontario, it took only a moment—the specific year was 1919—before they appeared just to the south, in upstate New York. Between then and the 1970s, coyotes followed two migration paths east. One version of a coyote Overland Trail was in the north. Coyotes that got to upstate New York then moved southward down the mountain chains all the way to Virginia and North Carolina. Meanwhile, a southern migration route took coyotes through Louisiana and across the lower South. Reflecting a sense of confusion, American books on mammal distribution in the 1950s began to note coyote appearances not just in the Midwest but in New England, the mid-Atlantic states, and both the upper and the Deep South. Some biologists convinced themselves the only possible explanation was that tourists returning from the West on Route 66 had bought coyote pups at Navajo roadside jewelry stands, then released them into the wilds of the East. Maybe it happened, but that wasn't the cause of the coyote's takeover of America. Coyote resolve was the cause, and by the late 1970s it had led them to colonize all of North America—they even swam cold Atlantic waters to Cape Cod. Unless they stowaway to Hawai'i, they colonized their final US state, Delaware, in 2010.

The modern coyote story has not just been about coyotes in states where no one would have imagined them a century ago. As we all realize now, coyotes were coming to live with us. They found the rewilding rural East and South appealing, to be sure, but it quickly became apparent that individuals, pairs, and even packs were also setting themselves up in towns and cities across the country. The urban coyote phenomenon happened first in places like California, Arizona, New Mexico, and Colorado, where we had founded major cities such as Los Angeles, San Diego, Phoenix, Tucson, Albuquerque, and Denver atop existing coyote habitats. Their colonization of our cities, from small burgs like my hometown in Louisiana to the biggest, loudest, most frenetic of our metropolises, has become the wildlife story of our time. It deserves some explanation.

The starting point is this: the truth is that coyotes have never been solely wilderness creatures. The news media have given us a false impression that coyotes have no business in places like Los Angeles or Chicago or Manhattan, that for reasons related to either the inviolable nature of modern cities or the coyote's suspect character, coyotes in cities make up a bizarre and inexplicable invasion. Yet the

Introduction: American Avatar 9

Aztec rendering of coyotl.
From The Voice of the Coyote *by J. Frank Dobie.*
© *1947, 1949 by J. Frank Dobie, renewed* © *1976 by The Capital National Bank. Used by permission of Little, Brown and Company, Inc.*

archeological and historical evidence is undeniable: for the 15,000 years since we humans have been in North America, coyotes have always been capable of living among us. Something about our lifestyle has always drawn coyotes to human camps, villages, and cities. That something is ecology at its simplest, even if it makes us squirm a bit. A coyote's primary prey happens to be our close fellow travelers, the mice and rats that flourish around and among us in profusion. As for fearing us too much to tolerate our presence, coyotes have taken our measure far too perceptively for that.

So the urban coyote is not a new thing. By the time Europeans got to America, coyotes had long since sought out the major Indian cities of Mesoamerica. One of their initial experiments with urban living seems to have been in the Aztec city of Tenochtitlan, on the site that became modern Mexico City. Half a millennium ago, coyotes and city-dwelling Aztecs came to know one another very well. We believe this because of place-names within the city. *Coyoacan,* or "place of the coyotes," was the name of both a suburb of the Aztec capital and a religious cult devoted to coyotes. A thousand years later we still use a form of the original Aztec name for this small American wolf: *coyotl*, pronounced *COY-yoht*, accent on the first syllable, silent *l*, in their Nahuatl language.

The Aztecs were fascinated with animals that hunted, and one that hunted within their cities drew special attention. Their rich mythology produced numerous coyote gods, including one called Coyotlinauatl, ceremonies for whom featured acolytes costumed with tails, sharp snouts, and erect ears. The Aztecs associated another of their coyote namesakes, Nezahualcoyotl, with sensuous pleasure and thought of him as a patron of the high arts, naturally of song but also of poetry, a musical language. Another of their gods was Coyotlinahual, a coyote sorcerer known for his ability to assume the shape and form of people. The Aztecs also honored a deity they called Huehuecoyotl, or "Venerable Old Coyote," who sounds so much like the widespread North American god-avatar often called "Old Man Coyote" that the empire-minded Aztecs may have borrowed him from tribes far northward, in what is now the western United States.

Tenochtitlan wasn't alone in having an urban coyote population in Indian America. Coyotes seem to have been actively working at least on the margins of many Indian cities, villages, and camps in the centuries before Columbus. A thousand years ago, Chaco City, within present-day Chaco Culture National Historic Park in northwestern New Mexico, was the Vatican of a far-flung civilization in the deserts of the American Southwest. During times of religious gatherings it was the largest Indian city anywhere in the future United States, with a population in excess of 40,000. Chaco unquestionably had coyotes in town; coyote bones are common in the archeological sites of the inner city.

Until dogcatchers, dog pounds, and leash laws (as I detail at some length later in this book) began to curb impressive stray dog populations in the cities of the late-nineteenth-century United States, coyotes had a difficult time infiltrating American cities. They were always on the margins though. As early as the 1830s, in the mission towns of California's Central Coast, naturalist Thomas Nuttall described coyotes "tame as dogs" yapping every night "through the villages" of the region. But in post "dog war" America, coyotes began to trickle into urban cores. They first attracted modern attention in the cities of Southern California and Arizona early in the twentieth century. In Denver coyotes had become an urban presence by the 1970s. Chicago, in the 1990s, was next, and by roughly 2000 almost every city in the United States and Canada, no matter how small and picturesque or sprawling and ear splitting, possessed a thriving population of coyotes as full-time residents. Urban North Americans, who thirty years before had assumed that nothing wilder than starlings and English sparrows coinhabited their concrete world, suddenly discovered that a small wolf (if indeed they knew what a coyote was), thought of as an animal of the desert wilderness, was now trotting down their streets and through their backyards.

Maybe it really was "the end of civilization," as a shocked Manhattanite phrased it, but eighty years after wildlife managers confirmed the first coyotes in upstate New York, in 1999 strollers spotted a particularly adventurous coyote ("wild and unleashed!" according to one metro headline) loping through New York City's Central Park. "Otis" was a ballsy young male who had somehow followed rail lines and crossed bridges into the midst of the largest, most densely developed urban complex in the United States. Captured among skyscrapers that rivaled the red-rock canyons of the desert Southwest, Otis spent the remainder of his years in a zoo in Queens, an outcome somewhat reminiscent of the fates that befell some of the Indian performers from *Buffalo Bill's Wild West*, left behind to be pursued and ogled in the East in the 1890s.

Otis turns out not to have been the first coyote to venture into the Big Apple. In Van Cortland Park in the Bronx there is a bronze statue of a twenty-nine-pound female coyote, run over on an expressway nearby in 1995 and said to be the first coyote in New York City since a coyote visitor in 1946. But Otis seemed different, more settler than explorer. He was the advance wave of a coyote river that now joins the boroughs of New York with neighborhoods in Los Angeles, San Francisco, Seattle, Phoenix, Denver, and Chicago in harboring resident wild canids as part of big-city ecology in modern America. In 2006 another coyote, this one nicknamed "Hal," navigated his way into Central Park. New Yorkers' response to another coyote in their midst involved more than giving it a name. Hal adroitly evaded capture long enough to trail a shouting entourage of policemen, park workers, reporters, photographers, news helicopters, and assorted street people. A spellbound media followed his adventures for days before Hal fell to a tranquilizer dart. Speculation was that he had gotten to Midtown from Westchester County via an Amtrak train bridge over the Harlem River.

Tragically, Hal the coyote died a few days after capture, probably from the stress of days of harrowing pursuit through downtown Manhattan, although he did also suffer from heartworms. But four years later, in 2010, another coyote, this one a female, got two months of freedom in Central Park before her capture. That same year also brought news of another female in Harlem, of one in Chelsea, of a coyote hit by a car on the West Side Highway, of three coyotes spotted on the grounds of Columbia University, and of coyotes successfully raising litters in Van Cortland Park and cavorting on a golf course in the Bronx. In 2011 residents of Queens began phoning in frequent coyote sightings. In 2012 one crossed a bridge onto Staten Island—the island's first coyote—and in 2013 the New York Police Department (again trailed by a swarm of reporters) pursued and darted a coyote in Crotona Park in the Bronx. In the spring of 2015, a coyote once again captured a news cycle in the Big Apple by nonchalantly peering down at city traffic from the rooftop of a bar in Queens, then doing a Hollywood action hero escape through the broken window of a nearby building. By that point the coyote population in the city was robust enough that biologists Chris Nagy and Mark Weckel were already seven years into a study of the animals in their inner-city habitat, an endeavor they styled the Gotham Coyote Project.

If one's argument for civilization holds that wild predators should never roam in broad daylight through the boroughs of America's largest, loudest, most radically urban metropolis, then, truly, the end of civilization had arrived on paw prints in the snow.

What makes predators unusually compelling for so many of us lies deeply enough in the human psyche that it could be called a genetic memory. We identify with them because we, too, emerged out of the dim, hazy consciousness of our early origins to find ourselves fellow carnivores and pursuers of prey. But we also preserve more chilling memories, of the fitful night and the leopard, of bright teeth and being hunted down ourselves. To confront a predator is to stand before the dual-faced god from our deep past. That is why we look longer, more intently, with more studied fascination at predators than at other kinds of animals.

The fact that coyotes have now become the most common large wild predators most Americans have ever seen may be one reason why in our own time just about everyone has a coyote story. The tawny, tail-swishing, sharp-nosed wild dog of the American deserts is now our furtive alley predator everywhere from Miami to Anchorage, San Diego to Maine, and the stories are piling up. During a 2007 heat wave, a coyote strolls in broad daylight into a Quiznos sandwich shop in Chicago and hops up on a freezer to cool off. Customers and staff flee for the street, where a shocked crowd gathers to peer in the windows as the coyote commandeers the store. In Los Angeles in 2009 a coyote snatches Daisy, actress Jessica Simpson's beloved Maltipoo, almost from her grasp. The tearful actress offers a reward for her pup, but Daisy, like a fair number of pets in Coyote America, is never heard from again. Meanwhile, the Internet is abuzz for a few days in 2012 with another story (it turns out to be true) from the Pacific side of the country. A California couple, cruising along at freeway speeds in the early morning on the Utah-Nevada border, drive through a pack of coyotes crossing I-80. Six hundred miles and ten hours later, unpacking the car in their driveway near Nevada City, they discover a full-grown coyote snagged like a bug in the grill of the car. Their flying coyote ornament is fully alert, with one cut on a paw and another on its muzzle. After hitchhiking from Utah to the West Coast, it is otherwise unhurt.

This twenty-first-century American familiarity with the coyote is simultaneously a new thing under the sun and—if you grasp something of the wholly remarkable history behind it—a revelation, about not just one animal but two: them and us. Like all species, coyotes have a history. It's just that not many animals on any continent have a history that even comes close to the one they have managed to fashion. We're one of their few rivals in biography; it's another of our many convergences.

"Coyote America" is what I call their story. Understanding the twists and turns of it, the historical roots and the modern scientific sense of an animal that is demonstrably under no one's control but its own, can help explain why coyotes are enveloping us. But naturally, in more ways than you would imagine, this story is about us. The coyote is a kind of special Darwinian mirror, reflecting back insights about ourselves as fellow mammals. Europeans had old experiences, stories, myths, and preconceptions about gray wolves, bears, and foxes and long employed folk tales about them to investigate human nature. But coyotes are different. The coyote is an American original whose evolutionary history has taken place on this continent, not in the Old World. We see it not from the traditional vantage but from a sideways one, and from that perspective everything looks different.

The first human eyes to assess the coyote for clues about its nature, then to be struck by what those clues revealed about the observers themselves, belonged to Indian peoples more than one hundred centuries ago. Only in the past five hundred years have Americans of European, Asian, and other backgrounds tried to make sense of the coyotes of the continent. So what the coyote offers up as illumination about us is often America-specific. These insights are edgier than the kinds of truths in animal tales from Europe. They are more ironic and growing hipper all the time. Some thoughtful Indian observer many thousands of years ago took precedence in this epiphany, no doubt, but it turns out the coyote really is The Dude, and The Dude *absolutely* abides.

Coyotes initially confused the Europeans and Africans who arrived in North America. What kind of animal was this insolent wild canine? A wolf? A fox? A jackal perhaps? After a couple or three centuries of indecision, we ended up making a conscious (and even datable) effort to malign these creatures, a sentiment that gained speed in the twentieth century, until we were hurling around wild epithets like "Original Bolsheviks." Eventually no American animal called up more heroic measures from us to achieve its total eradication. Yet for reasons that mystified everyone until just a few decades ago, coyotes proved they could take every haymaker we could throw, then respond with an almost nonchalant takeover of the ground we were standing on. By the time our attitudes about them finally began to shift in the 1960s, coyotes were well into their own manifest destiny, even into a variation on the classic American melting pot.

Circulating among us now like ghosts of the continent's ancient past, as if to make us cognizant that we are new and barely real here, coyotes oddly appear to grasp with those vivid yellow eyes that they function as avatars, stand-ins to help humans see themselves. If so, they have often reflected an edginess back to us. Like us, coyotes are bold, sometimes aggressive, occasionally menacing. Edgy. We certainly see that in them and perhaps perceive it in ourselves. No better example of edginess in the coyote-human relationship exists than this: as a general topic of conversation in today's America, coyotes are political. That shouldn't really be surprising, on reflection, but coyotes have a remarkable ability to demonstrate just how easily we can find things to disagree about in modern society.

In the 1980s a Yale University study of the US public's appreciation of wild creatures offered dramatic proof of the success of a century of hate messages about coyotes. Yale's poll ranked coyotes dead last in public appeal—behind rattlesnakes, skunks, vultures, rats, and cockroaches. Then, as a result of a new predator appreciation that elevated gray wolves to media and environmental stardom in the 1990s, the coyote's stock as a smaller wolf cousin began to rise, at least among liberals and environmentalists. Coyotes have never risen to full gray wolf status as environmental darlings. But their colonization of our cities is, if anything, exposing humans and coyotes to one another with an intimacy that's allowing new generations of Americans to form their own opinions. The shock of that intimacy has earned them enemies but also an awful lot of admirers.

Coyotes may now have more fans in the United States (and, as an iconic American animal, around the world) than ever before, but in contemporary America, coyotes still do one thing more than anything else: die, at a rate unmatched by any other large animal. Other than a federal poison ban riddled with loopholes and a handful of state restrictions against leghold traps and coyote-hunting contests, coyotes enjoy no governmental protection against being killed. The best guess is that altogether we kill about 500,000 of them a year. Roughly once every minute, about the time it takes to read this page, someone somewhere is ending the life of a coyote. That is only the current body count in a history that easily wins them the title of most persecuted large mammal in American history.

Persecuting an animal in a battle you can't win is an act of political ideology. Indeed, the political thing with coyotes reaches surprising dimensions. It's hard to escape a sense that coyotes have joined religion, the Iraq War, Obamacare, and climate change as one more thing the culture warriors in America have to disagree about. Asking people what they think about coyotes is akin to asking them what they think of John Wayne. The answer is immediately diagnostic of a whole range of belief systems and values.

The political disagreement even extends to how to pronounce the name of the animal. Simple pronunciation, I've come to realize, can serve as a clue in coyote politics, if not a hard-and-fast rule. Defenders and supporters of coyotes, usually (but not always) from educated, urban backgrounds, tend to pronounce the animal's name *ki-YOH-tee*, with the accent on the second syllable and a *t* so soft it's almost a *d*. Americans from rural backgrounds—who commonly fill the ranks of those who manage coyotes, shoot and trap them, or fear them and want them killed—struggle with a three-syllable name, a rendering that apparently sounds pretentious. Or maybe as a word out of the Southwest, *ki-yoh-tee* just sounds too fancy, perhaps too Spanish. In any case, to rural people the comfortable name is *ki-yote*, accent on the first of the two syllables. (I make an effort to untangle this pronunciation war later in the book.)

Let me tell you what this simple difference in pronunciation can mean. Working on this book I had occasion to do a couple of public talks that left me with, let us say, an enhanced perspective on coyotes as a gateway into the culture wars. That began when, in 2013, a representative of a Nebraska society dedicated to the career and literature of a famous Great Plains woman writer invited me to do a talk in her honor. Asked by a board member what I had in mind as a topic, I innocently offered up "something on coyotes." There was a moment of silence, and then the midwestern voice on the other end of the line said flatly, "Let me check with the group on that," followed by this polite warning: "But if we have you do that, could you please call them *ki-yotes*?"

In fact the board said no, so I ended up doing a talk on a different topic that had nothing to do with coyotes. But apparently because I had merely suggested coyotes as a topic, at the dinner before the public event, not a single member of the board of this literary society, all of them well-heeled ranchers judging by their attire, bothered to greet or shake hands with their speaker. It was not one of Nebraska's finer moments.

A month later I was addressing a large and enthusiastic crowd in Northern California, this time doing a talk about coyotes and this time among obvious coyote fanciers. Inside a classic old theater where Mark Twain and Jack London had once held forth, the coyote's intelligence and life story were held in high regard. Everyone in the audience had a personal coyote story. Outside, the sweet and unmistakable scent of the marijuana harvest in the surrounding hills marked the time of year and suffused the balmy California air. Somehow that seemed appropriate as another coyote culture war marker; in California I was obviously speaking (as a friend put the matter) to "coyote-loving hippies."

As coyotes have moved in to live among us, denigration of their survivability has become common in conversation: "When the world ends, only cockroaches, rats, and coyotes will be left." What I'm about to say next will not fall sweetly on the ears of those who subscribe to that view. But here it is: few animals in the world come closer to mimicking us and our own unique abilities. I think the Indians' grasp of that is the reason, perhaps as distantly as 10,000 years ago, they designated an avatar form of the coyote as a principal deity, responsible for creating all of North America and for instructing them about the human condition. The resulting Old Man Coyote character is not only one of the oldest gods of which we have record but one of the most intriguing in all history.

In Far Eastern religions the term "avatar" refers to the earthly representation of a deity, but here I intend it to mean something closer to its use in modern computing, where an avatar is the graphic representation of the user in the cyber world, an alter ego or stand-in. To understand how coyote history mirrors human history enough that this use of "avatar" seems appropriate, one needs a grasp of the outlines of both human evolutionary history and the history of coyotes in America. Providing an account of those intersections is one goal of this book. There are more of them than you think.

It is human hubris to think of coyotes as instinct writ on the world. In fact, exactly like ours, their personalities are the products of culture and genes and how environmental influences turn those genes on and off. Despite our being social species, these influences spin off markedly different individuals among both coyotes and us. Yet one important similarity we and coyotes share across our long histories is a reliance on problem-solving intelligence for success, a badge of real distinction, it seems to me. Perhaps most tellingly, we two species share the trait of being successful across times of great change, an uncommon situation for many other forms of life on Earth.

What conclusions might we draw from the coyote as stand-in avatar for our own deep time history? I spin off one or two of my own ruminations about that elsewhere in the book, but I think I would rather readers form their own impressions as they track through the coyote's story. Suffice it to say here that as we humans head off into an uncertain and probably dangerous future of our own making, it might be wise to keep an eye on them. I, for one, am going to be very interested in how coyotes cope with the twenty-first century and what insights we might draw about our own circumstances from a coyote history that so often seems to mirror ours.

This book is in most respects a coyote biography. More than half a century ago, Texas literary giant J. Frank Dobie wrote a wonderful volume of coyote folklore, and about the same time legendary federal coyote nemesis Stanley Young characterized the animal as the Biological Survey's most frustrating opponent, an enemy members of the agency actually lumped in with fascism as a threat to the American way. Several contemporary writers, among them Barry Lopez, have written books redacting the fascinating literature of the Old Man Coyote stories into modern prose, a strategy I employ here in a more limited way. But I attempt something different from all these: to tell the story of this American original from the evolution of the canid family in North America 5 million years ago down to the coyote's present incarnation as the wolf we tried to erase but that instead ended up in our backyards. That story follows a long and winding road from then till now.

In our time it is difficult to imagine an American politician invoking a native North American animal to describe the American character. Totem animals—despite the Pacific Northwest's embrace of the salmon as a kind of regional icon—are not exactly de rigueur in the modern world. But the truth is, the coyote's unique history and similarities to us do appear to make it, by this point in US history, a damned fine candidate as a national totem. Many contemporary Americans may be too far gone into the Anthropocene to learn much from an animal nowadays, but Marc Bekoff, a professor emeritus of ecology and evolutionary biology at the University of Colorado who has worked extensively on coyotes himself, believes modern Americans "are really craving to be 're-wilded.' They're craving to be reconnected to nature."

Coyotes have played this role before, after all, once for western Indians, more recently for West Coast bohemians in the twentieth century. In twenty-first-century America there are certainly enough of us immersed in science, evolution, and the natural world to comprise a likely group of national coyote acolytes. So I hope at least that seeing coyotes regularly in the world and knowing something about the trajectory of their zigs and zags through history and how their story jibes with our own may move some to see something connate and sympathetic in this small wolf, which at this very moment is taking up residence within a mile of every person in the United States whether he or she understands its biography or not.

That said, I will borrow a phrase from Mark Twain, who turns out to be a central figure in the story to come. I am about to tell you a tale that has in it a touch of pathos.

CHAPTER 1

Old Man America

In the remotest time of early North America, after he had molded mud from the ocean bottom into mountains, plains, and forests to create the essential topography of the continent, Coyote was going along. He had placed stars in the sky, some as pictures, some as a latticed road across the night, some tossed willy-nilly into the inky black. He had arranged the year into four seasons, and he had populated the world with humans. As the special helper of the Creator, who seemed not especially interested in any of this hands-on creation work himself, Coyote had killed monster after monster on behalf of his human charges, whom he'd then located in good, monster-free spots across America. He had released animals like buffalo from underground and—admittedly, with a few unlucky mistakes—placed salmon and other fish in many of the rivers. He had invented penises and vaginas and taught humans what to do with them. The first technology, in the form of fire, came from Coyote. Then, not without some remorse, he had introduced death into the world.

Now, with all these fundamental creations in place, Coyote had no intention of stepping into the background or hiding himself. He wanted to enjoy how much humans appreciated his creativity. And he especially wanted to see how quick-witted they might be when he offered them up some grand illustration of their own nature.

One morning Coyote was going along and spotted a handsome young warrior who told Coyote he was embarking on a journey of war against his enemies. Although Coyote was actually a peaceful sort who thought war and battles to the death were very bad ideas, he told his new companion that he was a famous warrior and would be indispensable on the quest.

That first night, the warrior said, they would camp at a place called Scalped Man by the Fire. Coyote did not like the sound of that, but he went along. At the camp Coyote relaxed while the warrior cooked and did all the chores. Then Coyote took the best pieces of the meal for himself, even laying extra meat over his chest and legs in case he awoke hungry during the night. Sometime in the night Coyote heard a sound, and when he looked, there was Scalped Man standing over him. Quick as he could, Coyote swung his club, but somehow he hit his own knee, which caused him to yowl in pain, waking the warrior. "I have taken care of Scalped Man," Coyote told him, and they both went back to sleep.

Having clubbed his knee badly, Coyote limped through much of the next day but made it OK to a camp called Cooked Meat Flying All Around, which sounded more like it. But that evening, dining on the chunks of meat whizzing all around, Coyote heard the warrior describe the next night's camp, Where the Arrows Fly Around. Suddenly his knee took a turn for the worse. Coyote lagged far behind that next day, hoping to camp somewhere else, but the warrior led them on. That night arrows began to fly from every direction. The warrior stood and caught one after another, while Coyote twisted and twirled and crawled on the ground trying to avoid them, until one arrow grazed his arm. I have been killed, Coyote shouted. But when the warrior pulled him to his feet and he found himself still alive, Coyote asserted that actually his hurt knee had caused him to fall asleep, and he had been dreaming.

The next night they would camp at Where the Women Visit the Men. This sounded like an excellent camp to Coyote. His knee improved so remarkably that day that he got far ahead in their march. That night, after much fidgeting and anticipation on Coyote's part, a woman did come to him, but in the darkness he decided she was an old crone. Hoping for a much younger woman, he sent her away, only to see in the firelight as she turned from him that in fact she was young and very beautiful. Coyote cried out for her to return, claiming it had been some spirit who had told her to leave, but she vanished into the night.

The next camp was called War Clubs Flying Around. All that day Coyote's knee hurt so much that he barely managed to arrive at the spot. Sure enough, that night clubs hurled at them from every direction. The warrior caught two, one for each of them, but Coyote dodged and weaved so much that a club finally beaned him. When he came to, Coyote told the warrior that in his boredom he had actually fallen asleep. That's why he had been lying so flat and still.

Then the warrior told Coyote that their next camp would be at a place called Vaginas Flying Around. Coyote's knee at once felt entirely well, and he was ready to depart then and there. He pleaded for more details, but the warrior fell asleep. Coyote sat by the fire all night thinking of vaginas and how many he might be able to carry with him. His knee now stronger than ever before in his life, Coyote left early and ranged far ahead the next day.

That night, as promised, vaginas began to sail into camp, and Coyote could tell they were just the kind he liked, very young and very plump. For most of the night, juicy vaginas sailed by, maddeningly out of reach, with Coyote flailing and chasing and panting until he was near collapse. Finally, near dawn, Coyote caught one. But exhausted as he was, when he finally pinned and mounted it, his organ refused to rise to the occasion.

The next night they would reach their final camp, and the warrior told Coyote this one was called Where the Enemy Attacks. Without delay Coyote's knee began to throb, and all day he hung back on the trail, crying piteously. And sure enough, when the next morning came, enemies attacked from all sides. Coyote at once ran for far horizons but was overtaken, clubbed, and scalped. Meanwhile the warrior subdued all his enemies, then looked for Coyote.

When he knew all was clear, Coyote stood and announced that he was going along now, but the warrior should consider himself lucky that he *had happened upon Coyote. Otherwise he would have had to engage in this adventure with no help at all from a famous warrior.*

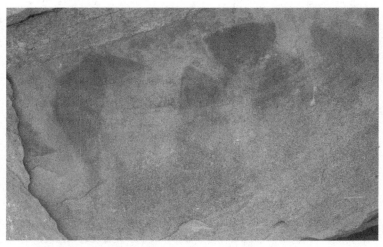

Indian rock art in Palo Duro Canyon, Texas, showing Coyote (left) and other characters from an Old Man Coyote story.
Courtesy Dan Flores.

Stories about Coyote or sometimes Old Man Coyote—and rarely about Old Woman Coyote, although they are present in the canon—are the oldest preserved human stories from North America. The truth is that Coyote (capitalized to distinguish the deity from the ordinary coyote trotting by while you read) is the most ancient god figure of which we have record on this continent. When Siberian hunters first started crossing Beringia or boating down the coastline 15,000 years or more ago, at some point in their entry of northwestern America they began to encounter coyotes for the first time. Wolves they knew from Asia, and well enough that at some point in their migration, these first Americans arrived with domesticated ones, wolflike dogs whose wild ancestors in those times were recent. But at least by the time of the Clovis people, who spread across the grasslands of interior America from Canada to Texas more than 13,000 years ago, continental coyotes were familiar creatures, and something about them resonated.

Religious explanations for the world and how it works are at least 40,000 to 50,000 years old, so these former Siberians no doubt arrived with intact religions, mythologies, and deities. But as these first Americans settled the part of the continent that would stretch from today's California all the way to the Mississippi River, from the Pacific Northwest to the future New Mexico and Arizona, Coyote emerged as the deity of the ancient continent. No one knows when this happened or exactly how Coyote became a principal figure in so many different peoples' creation stories and ruminations on the human condition. We know only, based on the oral Coyote stories collected among American Indians and set down by nineteenth- and twentieth-century ethnographers, that over the centuries the various tribes fashioned many hundreds of Coyote tales. No other native deity in America came anywhere close to inspiring such a vast body of oral literature. The story "Coyote and His Knee" is from the Wichitas of the Southern Plains, but I distilled the opening paragraphs of this chapter from groups as geographically separated as the Navajos (Southwest), Crows (Northern Plains), Karok and Wasco (California), Menominee (Great Lakes), Colville and Klamath (Pacific Northwest), and Salish and Blackfeet (Northern Rockies).

West of the Mississippi, across the last 10,000 years, Coyote has been America's universal deity, surviving as a Paleolithic god among agricultural peoples like the Wichitas and ultimately reaching as far south as the Aztecs, who knew him as Huehuecoyotl, Old Man Coyote. Or Old Man America.

For millions of years the grand expanse of the American Great Plains, extending northward to the boreal forests of Canada and southward to the deserts of the Southwest, was the biological Eden of North America, the continental version of Africa's Serengeti or Maasai Mara. Today, if you can find a piece of native prairie somewhere on the Great Plains that's away from the sounds of interstate traffic and beyond the stench of hog farms—anywhere will do, from Montana to West Texas—and if you're good at opening your mind to the possibilities of deep history, a few moments of imagining can bring this landscape back to life. A hundred centuries ago, elephants and camels and lions could have been in view. For thousands of years after that, herds of buffalo and horses—likely trailed by wolf packs and bands of native hunters tacking across the grass ocean and navigating by the Pole Star—would have grunted and grazed past your spot like wildebeests and zebras on the African veld.

And right in the mix of this wild, Africa-like bestiary of the Pleistocene were ancient coyotes, trotting in the midst of the ungulate herds, competing with other predators—and scavengers—for a living in the kind of world that would, indeed, have been familiar to our own ancestors on another continent halfway around the world.

As a singular animal emerging from earlier evolutionary canid ancestors, the coyote is a relative youth. Coyotes share evolutionary youthfulness with us. We are also young, our genus, *Homo*, emerging between 2.8 and 3 million years ago and our species coming out of its own "hominin soup" in Africa fewer than 200,000 years ago. The Canidae family appeared at about the same time, 5 to 6 million years ago, but halfway around the world, in North America, with some of its species beginning to spread out across the globe soon after. The ancestors of the gray wolf (*Canis lupus*), as we will see, became cosmopolitan and eventually colonized almost the entire planet, continuing to evolve before returning, group after group, to their natal American homeland.

Coyotes, it turns out, are also a kind of wolf. They shared a common ancestor with gray wolves down to about 3.2 million years ago, when coyote and gray wolf ancestors began to separate, first geographically, then, as distance increased, genetically. Genetic research indicates that today there is about a 4 percent genetic difference between coyotes and gray wolves. For perspective, that's roughly the same genetic distance as between modern humans and orangutans.

The histories of coyotes and humans have many parallels, but one difference is that across our own evolutionary history, we humans have created thousands of philosophies of meaning we call religions, while coyotes, so far as we can tell, embrace no religious tradition beyond *being alive*, sacred existence. Religions that feature animals as deities are probably the oldest forms of our own religious explanations; they are a type of religion called "animism," fashioned by humans living their lives as hunters or hunter-gatherers. What we might call "Coyotism" is, in other words, a Paleolithic religion. The famed psychologist Carl Jung is only one among hundreds of individuals, from scientists to poets, who have found the Coyote deity enduringly fascinating in part because of how fundamental he is in human thought. In Jung's view, Coyote is "a faithful copy of an absolutely undifferentiated human consciousness . . . a forerunner of the savior, and like him, God, man, and animal at once. He is both subhuman and superhuman, a bestial and divine being."

The Western religious traditions of Judaism, Islam, and Christianity sprang from later periods of human history following the domestication of plants and animals, which anthropologists call the Neolithic Revolution. Early Neolithic religions could feature animals—particularly the sacred bull—as deities. But over time, herding and agricultural cultures gradually replaced animal gods (along with gods of special places in the landscape, another feature of animism) with deities that assumed human form. The Greek gods, so foundational in Western cultures, are classic examples of this evolution. The Greeks replaced animal and plant deities with anthropomorphized gods and goddesses 4,000 years ago: Artemis became a "mistress of the animals" as goddess of the hunt, and Demeter evolved into a human-form goddess of wheat and crops. Coyote himself, it turns out, made at least a partial transmogrification toward human form.

One of the most intriguing questions about Coyote is this: Why did these first, ancient settlers of North America pick this particular animal as their deity? Ten millennia ago the first Americans would have had many scores of animal candidates for their deity figures. Charismatic creatures like mammoths or dire wolves or saber-tooth cats might seem to us more likely choices, and in the early stages of human settlement, perhaps they had been gods. I speculate that as the Wisconsin Ice Age gave way to a rapidly warming world, coupled with the great simplifying event known as the Pleistocene Extinctions (which took all three of my suggested species and many others), wild coyotes captured the imaginations of the Indian peoples of the time as creatures endowed with special abilities. I suspect that the coyote's evident skill in surviving those profound changes, when the big, charismatic species could not, attracted human attention. An easy identification with the social lives of predatory wild coyotes also probably made them feel familiar to human hunters.

In *Pueblo Gods and Myths*, anthropologist Hamilton Tyler writes, "The ability of an animal to become a god is in part due to his symbolic potential; which is to say, the number of ideas he can stand for. . . . A god, even the simplest god, is based upon a certain amount of abstraction in the human mind." Another anthropologist, Lewis Hyde, writes in *Trickster Makes This World*, "Coyote stories point to coyotes to teach about the mind; the stories themselves look to predator-prey relationships for the birth of cunning." Hyde continues, "One reason native observers may have chosen coyote the animal to be Coyote the Trickster is that the former in fact does exhibit a great plasticity of behavior and is, therefore, a consummate survivor in a shifting world." In a world of giant animals whose fates seemed so mortal, coyotes were almost magical.

Especially before our lives in cities, which obscured our deep dependency on nature and diverted our powers of observation, we humans were profound observers of the natural world. Early Americans would not have failed to notice one other characteristic of wild coyotes in a dangerous and changing world: that their uncanny ability to survive everything nature threw at them lay in a remarkable intelligence. The trickster that Lewis Hyde mentions is a very old human religious figure; found in many animistic religions around the world, he takes the form of many creatures—hares, spiders, blue jays, ravens, even humans, like the Norse trickster Loki. But across America, the coyote took up the mantle in the critical formulation of a god who lived by his wits. Having a smart god, after all, was crucial to survival, as well as to penetrating human nature and the animal within.

Soon after coyotes came to the attention of Western science, a debate raged among naturalists about whether these animals were in fact a kind of American jackal, related to the side-striped, black-backed, and Simien jackals of Africa or the golden jackals of Africa/Eurasia. Plenty of people in the nineteenth century looked at coyotes and saw jackals. Some still do. Biologist friends in Yellowstone Park told me about an African biologist who visited recently and had little interest in the wolves. Instead, he wanted to see coyotes, and when he did, he cut straight to a suspicion he obviously already held: "That's a *jackal*," he told them.

However much jackals and coyotes satisfy the old naturalist-derived kinship models of close resemblance, modern molecular genetics shows, as mentioned above, that *some* jackals separated from the wolf line 5 million years ago, about the same time that humans and chimps were diverging from a common ancestor. The golden jackal is genetically distant from the coyote it so strikingly resembles, again, by about 4 percent. We humans are also distant from gibbons, another of our primate cousins, by about the same amount. Perhaps to other species, humans, orangutans, and gibbons are tough to tell apart too.

The method for determining kinship among species has changed drastically in recent decades. A major 2009 publication on the evolution of North American canids, produced by the American Museum

Golden jackal, the coyote's distant cousin in Africa,
southern Europe, and southern Asia.
Courtesy Wikimedia Commons.

of Natural History, represents the classic approach to animal evolutionary relationships, determining kinship based on examinations of fossil ancestors and morphological measurements. I spoke with one of the study's authors, Dr. Xiaoming Wang of the Los Angeles Museum of Natural History, who told me that the fossil evidence is clear that the Canidae family evolved in North America, probably in what is now the American Southwest, about 5.3 million years ago. Fossils indicate that wolves and coyotes indeed share a distant common ancestor. Dr. Wang posits this was a species paleontologists know as *Canis lepophagus*, a primitive Ice Age wolf that, like American horses, became geographically widespread by crossing the land bridges connecting America to Eurasia. Some of this primitive wolf's populations spread beyond America as early as 3.5 million years ago. Others stayed behind.

Continuing evolution of this particular canid line back home in North America led to a species that may be at the center of several current scientific debates about wolves and coyotes. *Canis edwardii* was a small wolf of American roots whose oldest fossilized remains come from 3-million-year-old Blancan Age sites scattered from California to Nebraska to West Texas. Maintaining a presence in North America for the next 2.5 million years—as late as the 1960s and 1970s, some wondered if it had ever gone extinct—*C. edwardii* was somewhere in size between a coyote and a gray wolf, although in early forms it lacked the sagittal crest that characterizes both wolves and coyotes today. But Wang and the paleontologists from the American Museum of Natural History are convinced that roughly 800,000 years ago, from some population of this early wolf, American environmental conditions began to select for smaller, quicker canids. But would these smaller, more refined successors replace the larger, wolfier ancestor in North America? Or would both survive?

We modern humans are fairly uncommon evolutionarily in having replaced all the progenitors in our line. But while we are the sole surviving species in our genus now, this was not always so. Modern humans evolved in Africa some 300,000 years ago, but especially once we spread out of Africa and into Europe and Asia 45,000 years ago, we had to share the world with two other species in the genus *Homo*. The Neanderthals and Denisovans had preceded us in leaving Africa and adapting to cold climates in Asia and Europe. So for at least a few thousand years—some argue for 15,000—down to some truly epic moment when the last Neanderthals passed, we lived alongside them. Probably, we also exterminated them. But most intriguingly of all, at least occasionally, individuals from these different groups of humans seduced one another. We preserve some of the genetic markers of those other genomes in our own, but we ended up as the only species of *Homo* left on Earth. That's not quite how things have played out so far in the genus *Canis*.

The wolfy ancestor, *C. edwardii*, produced several lines of smaller canids, and Wang's interpretation of the fossil evidence yields a better explanation for why some naturalists confused coyotes and jackals. Roughly 1 million years ago, a population of these smaller versions of *C. edwardii* migrated across a land bridge to the Old World and became *Canis aureus*, the golden jackal, the animal that had some early naturalists convinced that coyotes were American versions of this Eurasian/African jackal. Meanwhile, the related small *C. edwardii* offspring that remained and continued to evolve in North America became our own *Canis latrans* and other closely related forms. These early coyotes spread across America, with specimens appearing in the fossil record from California and Colorado to Pennsylvania and West Virginia.

In North America, events seem to have unfolded with coyotes and their *edwardii* ancestors something like they did with us and Neanderthals. As closely related species in the same genus, they coexisted—one a wolf, the offspring a wolflike coyote—in an America where the ebb and flow of glacial ages produced not just the rapid evolution and extinction of many species but land bridges that brought grand migrations of new creatures into America. As the climate swung wildly from icy to warm, habitats changed with dizzying speed. At places like the La Brea Tar Pits in Southern California, the fossil assemblages of the late Pleistocene, from 1 million down to about 11,000 years ago, show that *Canis edwardii* and *Canis latrans* were in fact present side by side at kill sites, much as modern humans and Neanderthals appear to have shared different parts of the same valleys in France and Germany during the last Ice Age.

But what if another hominid, or better, a couple of them, had joined us in the Gorges de l'Ardèche in southern France 40,000 years ago? In effect, that's what happened to the American canids in the late Pleistocene. For 3.5 million years the descendants of that original migrating American wolf, *Canis lepophagus*, had been evolving in Asia. Now, as the great beasts of the late Pleistocene—mammoths, mastodons, longhorned bison—migrated out of Siberia across the Bering land bridge to range among vast herds of horses and camels on the mid-latitude plains of North America, gray wolves followed them. A *lepophagus* descendant called *Canis chihliensis*, which lived in China around 2.7 million years ago, in the late Pliocene, seems to have been the progenitor of these newly migrating wolves. The first of its offspring to return to its American homeland was a wolf known to paleontologists as *Canis armbrusteri*. We know of this large wolf from fossils in the American Southwest dating to 2.5 million years ago, but it is most famous for begetting a species of giant wolves never forgotten by anyone who has ever read about them (or watched *Game of Thrones*). A quarter million years ago, *Canis dirus*, the gigantic dire wolf, seems to have emerged as a species on this Great Plains Eden of the Animals, where it joined American wolves and early coyotes in the hunt.

The world of American canids was about to get even more crowded because another large wolf stemming from the previous million years of Eurasian wolf evolution was coming. *Canis lupus*, the modern gray wolf, arrived in Europe (the famous "Wolf Event") 1 million years ago. Its late origins may have been in Siberia, where it probably evolved in response to the presence of so many large grazing prey animals. But just as they followed the great herds in Europe—all those magnificent creatures our modern human ancestors painted on the limestone walls of Chauvet Cave 30,000 years ago—gray wolves also accompanied similar herds that followed ice-free corridors from Siberia into America. The gray wolf, much changed from its travels abroad, was coming home.

Canis lupus, like us humans, was a relatively late arrival in America. Both Asian humans and Asian wolves entered North America only in the last 20,000 years. But once it joined the hunting and scavenging of other American canids in the grand predator picnic of the Pleistocene, the gray wolf decidedly made its presence felt. Like velociraptors, the meat-eating dinosaurs that filled a similar niche in the Americas 65 million years earlier, gray wolves formed packs to hunt medium-size herbivores like bison and elk. A modern American ecology began to take shape.

No doubt to everyone's relief, neither the frightening short-faced bear nor dire wolves the size of small horses were destined to survive the Pleistocene. Roughly 10,000 years ago, our last dramatic extinction event, the Pleistocene Extinctions, carried away many of the African-like giants that roamed America then. Late-arriving gray wolves survived, however, and so did America's own early forms of coyotes. But losing thirty-two genera of the most dramatic animals of the continent forever changed the world of American predators, one of them particularly.

The evolution of a predator has more to do with its prey base than with its competition with other predators, but at this point in the coyote's history, competition with wolves emerged as a powerful shaper. Studies of the fossils of Rancho La Brea Tar Pits show how this probably worked in coyote evolution. In the late Pleistocene, a coyote subspecies that paleobiologists call *Canis latrans orcutti* appeared most commonly at kill sites alongside gray wolves, dire wolves, and American wolves. This subspecies is intriguing for what its fossils say about its size, and presumably its niche, back then. *Orcutti* coyotes were almost wolf-sized, with far more massive heads and dentition than modern coyotes. As long as America's Pleistocene bestiary had remained intact, packs of *C. l. orcutti* had clearly prowled the kill sites, competing with various kinds of true wolves for the largesse of the Pleistocene.

But in the wake of the Pleistocene Extinctions, as large grazer after large grazer disappeared from North America, the coyotes of 10,000 years ago did something wolves did not: they shrank. As the extinctions significantly altered the remaining prey base, and presumably as competition between gray wolves and the strapping Pleistocene coyotes intensified, the adaptive genius of the coyote was to back away and seek out new, smaller prey, creatures that didn't have to be brought down by packs but could succumb to individual efforts. Wolves remained big, five- to six-foot-long pack hunters weighing 80 to 120 pounds. Coyotes shrank to become three- to four-foot-long, twenty-five- to forty-five-pound hunters of smaller game. They became scavengers, even omnivores. The most recent work on coyote evolution in this critical period, in an article in *The Proceedings of the National Academy of Sciences*, puts the matter directly: "Interactions

Coyote scat filled with juniper berries, part of coyotes' omnivorous diet and perhaps an astringent against worms.
Courtesy Dan Flores.

among carnivores and their prey are the probable cause of evolutionary change in coyotes."

This evolution conferred on the coyote a rare and remarkable adaptation, one familiar to anyone who knows the deep history of our own species. In the coyotes' case, this trait would go far in helping them become so unique, so resilient under mortality pressures, and so successful across the next hundred centuries of North American history. Coyotes share this adaptation with very few other species, but they do share it with us.

Like canids, all humans are intensely social. But we developed a unique kind of social life. The anthropological term for the flexibility we show as a social species is the same as the zoological term: "fission-fusion." Translated, that means that among a rare few species, evolutionary pressures have selected for sociality that allows for unusual flexibility among individuals, who can be either gregarious or solitary as conditions warrant. In sociality of this type, individuals cooperate with other members of their species to take on any manner of large endeavors. Most readily this occurs among closely related individuals, although some anthropologists believe that our next step—cooperation among nonrelated groups—enabled modern humans to colonize the planet. But species that show fission-fusion adaptations also allow for more individualized effort under other circumstances, a flexibility that can confer tremendous survival and colonizing advantages during disease outbreaks and other intense mortality events. Despite the myth of the "lone wolf," the gray wolf is not a fission-fusion carnivore; it evolved as a specialized pack animal to pursue large prey. Its strong pack ties and lack of social flexibility became near-fatal flaws among wolves in the modern age, when we learned to use their social instincts to trap and poison them. Their sociality, in fact, enabled us to extirpate wolves in the United States.

Most predators are either solitary or social, not both. But like us, the coyote gets to have it both ways. Real danger from larger gray wolves fashioned a plasticity in its behavior, giving it the flexibility to move freely between joining a collective and pursuing individual or pair strategies. As circumstances called for it, coyotes could become pack animals when catching prey like deer (or defending against wolves) called for cooperation. But coyotes could also function very well as solitary hunters, focusing on the kind of small fare—mice, voles, rabbits, even insects, and a wide array of vegetable foods such as juniper berries—that an individual animal could survive on. (While they have the teeth and jaws of a predator, coyotes also possess molars that enable them to grind and chew vegetables.) Less pack-adapted than wolves, coyotes even evince fewer facial expressions. For coyotes, fission-fusion created possibilities that allowed them to carve out an ecological niche in post-Pleistocene America different from those of both wolves and foxes. When, centuries later, twentieth-century American predator policy resolved on exterminating them, they used that biological flexibility to help thwart the effort.

But long before that, the genius of post-Pleistocene coyotes' lifestyles attracted the attention of America's first human inhabitants, who appear to have seen so much of themselves in the coyotes around them.

Coyote, in the earliest mythologies of North America, is not actually what one might call the Ultimate Cause god. More often, as in Coyote stories from people like the Salish and Nez Perce, he is the immortal right-hand helper of some abstract First Cause. His divinity is only semi, perhaps because he is present on Earth and engaged among humans. Most often in the stories, however, Coyote inhabits the world before humans are on the scene. Sometimes his initial form is human, which he gives up for his coyote body once humans are present. In stories set following his creation of the continent, though, Coyote is commonly a kind of anthropomorphic animal. In effect he is a Coyote Man. He preserves a tail, sharp muzzle, and erect ears, but he stands and walks upright, has a wife and family, and displays normal human fixations on status, food, fun, and lust. He is also capable of shape-shifting into a form so humanlike that often the other characters in a story only suspect, due to his behavior, that they are dealing with Coyote himself.

When one reads American Coyote stories, it does not take much time or analytic effort to conclude who Coyote really is, and it is that realization that makes him so intriguing as a god. Coyote is us in avatar form, or perhaps something more like The God Within. In the beginning myths the Coyote Man semideity's function is creation itself. Coyote takes the basic structure of the world as set in motion by the First Cause, then "improves" on it and gives it the natural laws that make it work. That done, his larger purpose in the many oral stories about him is to reveal the "man" component of his personality. Fascinating to me, unlike a perfect deity—such as Carl Jung's "savior" figure or a Jesus who teaches a codified morality and sets himself up as role model to humans striving for godlike perfection—Coyote personifies the full suite of humanity's traits. He is a god who is not merely good but also, transparently, very, very bad.

So as a literary character, Coyote is the full monty. He is no simple caricature but rather a complex figure full of nuances of all sorts. Coyote is admirable, inspirational, imaginative, and energetic—a whirlwind biophysical force with a large capacity for taking sensuous pleasure in life. But, as in the opening story, "Coyote and His Knee," he is also vain, deceitful, and ridiculously self-serving. And I should add that he is quite often envious, lustful to a degree of advanced, creative horniness, and possessed of an overconfidence that gets him into no end of fixes. Coyote's commonest flaw in the literature is probably a consequence of the way his human traits, both positive and negative, combine. That is to say, he finds reasons—sometimes because he's admirable; more often because he suffers from various forms of narcissism—never to be quite satisfied with the way things are. And because inevitably he is unable to predict consequences with any accuracy, his tinkering with the world usually produces disaster, especially for Coyote himself. The stories are funny because Coyote is a trickster who is forever falling for the oldest trick in the book.

As North America's oldest surviving deity, Coyote bequeaths to us down the timeline a continental world of imagination, creation, and artistry but also of self-absorption, hubris, and big trouble. As the stories seem to ask, how can one not see the Coyote impulse writ large in humanity? Indeed, given what we now know about ourselves, compliments of the fields of evolutionary psychology and neuroscience, when we look back at a Coyote canon that's thousands of years old, what does it say about how well Coyote—that is to say, people—grasped human nature long before modern science emerged to help us figure ourselves out?

One of the most ardent modern advocates of our coming to terms with human nature by examining our evolution has been Harvard biologist Edward O. Wilson. Back in 1978, when Wilson wrote in *On Human Nature* that "we are biological" and "we have no place to go but Earth," evolutionists had not really tackled the task of understanding the human condition. But Wilson has been a player in the field ever since. "There is no predestination, no unfathomed mystery of life," he writes in 2014's *The Meaning of Human Existence*. "Demons and gods do not vie for our allegiance. Instead, we are self-made, independent, alone, and fragile, a biological species adapted to live in a biological world."

As an early founder of evolutionary psychology, Wilson (and now many others) believes that much of the basis of human nature derives from our evolution as a social species, one that happens also to be omnivorous but for whom meat eating became an important cultural driver. Human evolutionary behaviors are often the most generalized and least rationalized of all our actions. And a universal example is our genetic instinct for self-absorbed, selfish personal behavior, as described in Richard Dawkins's famous "selfish gene" hypothesis. Another is a usually unquestioned bias toward youth and signals about health, cues to sexual fitness in a Darwinian world. These characteristics are so inherent that they are often invisible to us, yet direct much of how we act in the world.

Wilson and the evolutionary psychologists are not always optimists about what our evolution did for us. Our evolutionary adaptations allowed us to spread across and dominate the world, but unless we understand their effect on us, the same behaviors that made us so wildly successful can augment our inclination to engage in wars of tribal aggression over resources or to degrade the environment of the planet.

One of those adaptations, namely the fission-fusion trait we share with coyotes, may have also bequeathed to us one of the grand dilemmas of being human. According to a recent (and fascinating) idea in evolutionary psychology, natural selection among humans operates on two levels: that of individuals within social groups and that between competing groups. The intriguing strategy for survival we acquired early in our evolution effectively instilled in us directly contrary behaviors: both selfish and individualistic impulses and altruistic and cooperative ones. If mythic tales like the Coyote stories were perceptive enough to shine an ancient light on human verities, this grand conflict—the struggle inside the human mind as we attempt to sort out survival-dependent impulses to be both ignoble and noble—should be a common theme. Naturally, it is.

Modern neuroscience is another field that is offering up epiphany-like insights into how our brain's evolution and our internal chemistry have shaped us. An unusual historian of science whose work certainly intrigues me is Daniel Lord Smail. Smail has tried to illuminate how evolutionary neuroscience might be working on human behaviors we have all observed but been unable to explain. His book on the subject was 2008's *On Deep History and the Brain*. Old Man Coyote does not appear in it, at least not by name. But he is there, inhabiting the motives behind scientific terms that did not yet exist when he was serving as a Coyote Man avatar for ancient North America.

Smail is onto a phenomenon others might associate with the evolution of novelty-seeking genes. He avers that human evolution was never just about equipping us for survival in nature; we also evolved in response to our social lives in primate groups. A stimulating social life full of emotional highs and lows helped produce our neurochemistry of transmitters like dopamine and serotonin, chemical "washes" triggered by our experiences and sensations. Sometimes these washes became addictive enough that we deliberately sought experiences to induce them. Smail contends that the emotions of our evolutionary social lives quite literally set us up to seek out other neurotransmitter catalysts as we colonized around the world—psychotropics like coffee, sugar, tobacco, chocolate, opium, alcohol, cocaine, and marijuana. He rather mind-blowingly concludes that goosing our neurochemistry makes all kinds of spectacles, along with drugs, shopping, recreational sex, and pornography, pretty much predictable, given our evolutionary history as a social species.

A mind conflicted about "sin" and "virtue"; self-interest versus cooperation; games of love and status; new experiences to jolt neurochemistry and emotional states: for the long-ago Americans who selected a part man, part coyote as a suitable avatar for understanding themselves, what better subjects for the adventures of their Coyote god than these?

I think of America's Coyote stories as universal. Because they are in large measure commentaries on human nature, Coyote stories belong to all of us for the same reasons that Shakespeare's plays or Dostoyevsky's novels do. But of course they did emerge out of a cultural context—or contexts, more accurately, for different Coyote stories sprang from different peoples and take place in all manner of geographic settings. Many scores of generations of Indian peoples passed down the oldest, and the method of passage, until professional and amateur ethnologists and folklorists began to transcribe the tales a century ago, was always oral. For several thousands of years in America, Coyote stories came down the generations via spoken delivery, usually to mixed audiences of men, women, and children gathered in winter lodges lit and warmed by open fires. No doubt over the centuries storytellers of Mark Twain–like brilliance dazzled audiences late into the night with the many astounding adventures of Coyote, thereby romanticizing their peoples' trajectory through time.

Take the measure of this large literature, and it is evident that Coyote stories packed a cultural punch no media form today can really match. If we could somehow invent a modern art form that combined the religious/philosophical explanations of holy texts with the spectacle of modern movies, the entertainment value of video games, and the magic realism of Latin American novels, maybe it would rival the impact Coyote's adventures once had on audiences.

The stories' original functions were roughly sixfold. They provided explanations for why the world is the way it is: why a particular mountain range exists, why salmon are here and not there, why no one can return from death. All those things that seem unalterable about the world were Coyote's domain. Another function was educational. Coyote stories don't just offer up trenchant observations about human nature, pointing out our innate selfishness, our helpless striving for status, or our longing for sensations that trigger our feel-good neurochemistry. They tend to look at human foibles ironically or to make jokes about them while also showcasing proper behavior for a social species. Coyote stories also issued implicit warnings about threats to human survival. And they were wildly entertaining: it was (and still is) perversely pleasurable to observe a character who so blithely ignores rules and restrictions, usually with the predictable outcome, although Coyote benefits from rule breaking often enough to keep things interesting.

But what, no moral code in these stories? No promise of eternal life, no salvation from death, knowledge of which forms the ancient and oppressive burden of our self-awareness? Coyote stories offer up none of these things. They do proffer a firm nod to the "religion" of the wild coyote. Old Man America teaches delight in being alive in a world of wondrous possibilities.

It ought to be said, finally, that Coyote stories are not really for visionary dreamers who expect to change the world. Coyotism is a philosophy for the realists among us, those who can do a Cormac McCarthy–like appraisal of human motives but find a kind of chagrined humor in the act, who may think of the human story as cyclical, even predictable, because human nature never seems to change and is such a powerful presence in history. With modern scientific help in understanding why we are who we are and why certain aspects of life tempt us the way they do, it is fascinating to look at North America's Coyote stories to see how deeply into human time our observations about our foibles extend. Coyotism tells us that while we may long have misunderstood the motives for our behavior, we've also long known how human nature expresses itself. And who better to illustrate that than self-centered, gluttonous, carnal Coyote?

Coyote had spotted a most beautiful girl, a chief's daughter, and had decided she was the one for him. As beautiful chief's daughters do, she ignored him. Coyote, however, had heard that there were White Men along the eastern seashore who had many wonderful things. With magic, Coyote went there, and when he returned he brought with him four things no one had seen before. Then he set up a lodge right beside that of the beautiful young girl and commenced to work and pound away the night, making such a noise that the girl could not sleep.

The next day she sent a relative to find out why Coyote was making such a commotion. The emissary returned with the news that Coyote was making things, wonderful things. When she heard this, the chief's daughter was curious. So she and her relative met with Coyote, and the chief's daughter asked him what he was making. Coyote proudly showed her a bead choker made from colorful glass beads he had acquired from the Whites. The necklace was strange and novel. She wanted it badly.

"What do you want for this?" she asked.

"Nothing much, just a kiss," Coyote replied. The two women discussed this request, agreed there was little harm in it, and the first of their bargains was consummated.

That night Coyote set up a tremendous racket in his lodge, and again the beautiful girl wondered what he could be making. So the next day she asked, and he showed her a wonderful iron kettle, new and shiny and far better for cooking in than anything her people owned.

"What do you want for this?" she asked him.

"Oh, well, nothing much. I just want to fondle one of your breasts." It seemed such a small request, the two girls decided, so the deal was struck.

Throughout the third night Coyote thrashed and howled and set up a din in his lodge, and the next morning, when the question came, he showed the beautiful young girl a red wool blanket with stripes in several colors. It was the most wonderful thing she had ever seen, and this time, anticipating him, she offered him the chance to fondle her other breast in trade.

"No," Coyote said, "what I want for this blanket is to feel one of your buttocks." The girls consulted, and to possess this astonishing blanket, she permitted it.

That night Coyote outdid himself. The noise was deafening, more than all the rest of the nights combined. The next morning the chief's daughter was at his lodge very early, and Coyote showed her the night's result. It was a mirror, the first she had ever seen. She gazed for a long time at herself in its reflection, knowing in her heart she must have this new creation for her own, and then she asked the question Coyote was betting on.

"Oh, not much, really," said Coyote in response. "Just a look between your legs."

So the bargain was made. But this time when the deed was done, Coyote shook his head, "Oh my, oh my, too bad, too bad," he said. "Your winyan-shan is upside-down. It has to be remade. It can't stay like it is. What a pity!"

The girl went home and thought long about this and then resolved on what to do. If she truly did need remaking between her legs, who else to do it than he who made such wonderful and priceless new things at night?

"Go and fetch Coyote, and do it quickly," she told her relative.

In the hundreds of stories about him, Coyote is many things. Among peoples like the Pueblos of the Southwest, the Coyote figure is even a fool, the butt of jokes. But no matter what form he assumes, in many of the stories his character turns his pointed canine nose up at the proper social behavior—overcoming self-interest and acting unselfishly on behalf of others—and does exactly the opposite. Coyote, in other words, operates as the god of Dawkins's selfish gene, and in this form he takes on the character of a self-absorbed, narcissistic buffoon. The story's arc then holds such behavior up in plain view for comic ridicule.

But sometimes Coyote himself calls out selfishness.

One day, as always, Coyote was going along. This particular morning he happened upon the Frog People, who since the start of time had monopolized all the water. Anytime someone needed a bath or water for cooking or even a drink, he or she had to barter or beg for it from the Frog People. All the water was in their possession. So Coyote offered to pay them with a seashell he'd found so that he could take a deep drink from the water they had impounded behind an immense dam. Coyote drank and drank, and he drank for such a very long time that the Frog People finally grew suspicious. And they should have, for Coyote was not just drinking. He was also digging away intently at the base of the dam, which finally collapsed and released all the water behind it across the world.

The Frog People were furious. They had owned all the water. But Coyote shamed them. It was not proper for one group to hoard what everyone needed, he told them. Now water was available to all.

Coyote did this sort of thing too. He could also act in the interest of the group.

In their settings and perhaps in actual depth of time, Coyote stories about death appear to be an early, critical part of the Coyote genre, and to say that Coyote has a very complicated relationship with death is to state the obvious. As deities he and his friend Fox are immortals. Whenever one of them is killed, typically the other brings him back to life (although in such stories Coyote commonly insists he has "just been sleeping"). But stories from across western America lay death for all the rest of us directly at Coyote's doorstep. In tale after tale it is Coyote who decreed—for two admittedly admirable reasons—that all human beings must die. If humans never died, Coyote reasoned (in an explanation found in stories from both the Yanas of California and the Navajos of the Southwest), then overpopulation and the destruction of the Earth would result. Hence, the initial reason Coyote invented death was environmental: to save the world.

So it was Coyote who "made it law" (the Yanas said) that humans would have to die to create space for the generations down through time. When the first humans heard this, however, they resented it deeply. According to the Yanas, Coyote then came up with a second rationale for death, this time as the ultimate reason for appreciating being alive: "Well, you know, if you die, then you really have to take life seriously, you have to think about things more."

But it so happened that Coyote had a son who, running a race with humans, was bitten by a rattlesnake and died. Suddenly the tragedy of death affected not someone else but Coyote himself. Coyote set about wailing and gnashing his teeth, "dancing with grief" and "acting like a crazy man." But his son did not come back to life. Death was already in the world and could not be recalled merely because it had become personal for an avatar god.

When calamity visits him directly, Coyote has some serious second thoughts about what a good idea death is. A Nez Perce account called "Coyote and the Shadow People" tells one of the most poignant stories of Coyote's personal reaction to death. Set, in effect, in the days of Genesis, when only animals were on the Earth, it is something close to a North American version of the Greek myth of Orpheus and his slain wife, Eurydice.

Coyote and his wife were living happily when she became sick. When she died, Coyote was overcome with grief and loneliness. Others had died, but this was different. So when Death Spirit came to him and offered to take him to the place where his wife had gone, Coyote was filled with hope. "But, I tell you," said Death Spirit, "you must do everything exactly as I say; not once are you to disregard my commands and do something else."

So Coyote traveled with Death Spirit, thinking of his wife but noticing that his guide was very difficult to see and follow. He looked more like a shadow than anything real. When he pointed out herds of horses in the plain over which they traveled or bushes covered in serviceberries, Coyote saw nothing. But he exclaimed over the horses and pretended to eat the berries.

Soon enough the guide announced that they had arrived and led Coyote to where his wife was said to be sitting with many others inside a very, very long lodge. Again the Death Spirit cautioned Coyote to do exactly as he said. Coyote made every effort to do so, but while he felt the Spirit's presence, as far as he could see they were sitting in an open prairie. But Death Spirit told him that conditions were different here, that when night fell in the living world, it would be dawn in this place. Sure enough, when night fell Coyote began to hear people whispering. He began to see many fires in the lodge and to recognize old friends, whom he greeted and was able to walk about and reminisce with. And he was overjoyed to find his wife at his side.

After many wonderful hours, the beings in the lodge began to grow faint and hard to see. Then the Spirit came to Coyote and revealed that as dawn came in the living world, night came here. He told Coyote to remain where he sat and not move, and Coyote said he would. When dawn came Coyote found himself sitting in the open prairie. As instructed, he remained there all day, broiling in the heat but waiting motionless.

This went on for several dawns and several nights, with Coyote's friends and wife returning and making merry, then fading as dawn came, and Coyote waiting patiently in the heat of the day. Finally, after what seemed a very long absence, the Death Spirit came to him and said, "Tomorrow you will go home. You will take your wife with you." He told Coyote that they would travel for five days and pass five mountains, and while he could talk with his wife, under no circumstances should he touch her; he should never lay a hand on her until they had passed the last of the five mountains. Then the Spirit admonished Coyote, "You, Coyote, must guard against your inclination to do foolish things."

At dawn Coyote and his wife started out, although Coyote could barely discern her at his side. But when they crossed the first mountain, Coyote could feel her presence more strongly. When they camped on the homeward side of the second mountain, she became clearer to him, and at the next camp, beyond the third mountain, she became clearer still.

Now they were making their fourth camp, with only the final mountain to cross the next day, and Coyote could at last see his wife's face and her young body. She was almost a living being again. Coyote had dared not reach out to her before, but now, looking at her right there with him, he was overcome with joy at having her again and so impulsively ran to embrace her. "Stop! Stop! Coyote!" she cried. But it was too late. At the very instant he touched her body, she vanished.

On learning of Coyote's folly, Death Spirit was furious, and he did not hesitate. "You, Coyote, were about to establish the practice of returning *from death. Only a short time away the human race is coming, but you have spoiled everything and established for them death as it is."*

At this Coyote hung his head and wept. But then he had an idea. Drawing himself up, he retraced the journey he and Death Spirit had made. He tried with all his might to see the horses and taste the serviceberries. He found the spot where the long lodge had stood, even where he had sat with his wife beside him. And when night fell he strained to hear voices and see fires.

When dawn came, Coyote found himself sitting in an open, empty plain, all alone.

———

Ten thousand years ago, with North America drastically changing as a result of the end of the Wisconsin Ice Age and scores of its most characteristic species becoming extinct, the continent's indigenous wild coyotes had furnished a model of survivability for humans struggling through this epic environmental crisis. The somewhat shocking truth is that resulting stories about Coyote and the essence of his personality are probably older than the Gilgamesh epic at the foundations of Western civilization, and in the Americas they spread across as wide a geography. The Aztecs, 5,000 miles from Crow and Blackfeet country, preserved Coyote's likeness in their codices, stone effigies, and stelae. As befits Coyote's character in the stories, Aztec coyote cults in North America's most sophisticated civilization held festivals for him.

Gods come and go, but Old Man America was too useful a deity to abandon. As the Siouan "Winyan-shan Upside-Down" story testifies, Indians were still creating Coyote stories well after Europeans arrived, and they continued to tell them in their own communities into the twentieth century and beyond. Well beyond, at least long enough for American ethnographers and folklorists to discover their narrative richness and delights, as I discuss in the Epilogue.

Indian peoples also continued to find the wild coyotes that had inspired their doppelganger god of particular interest, worthy always of respect and often of veneration. Rarely did any Indian try to domesticate a coyote, though. Because of their evolution as smaller dogs in a wolf's world, domesticated coyote pups mature into nervous and highly unpredictable adults. Indians who lived among coyotes occasionally did breed their female dogs with both coyotes and wolves to restore some wild traits in their camp dogs. But unlike wolves, coyotes were never destined to become our domesticated familiars, which is just as well. To assign credit where it's due, coyotes did not choose domestication, thank you very much.

But more so than the host of other animals whose personalities, character traits, and special abilities Indians always studied closely, wild coyotes did appear to many native peoples as unusually powerful (and sometimes dangerous) fellow inhabitants of the continent. Barre Toelken has written of the Navajos of the Southwest that they drew no real distinction between the coyote in the desert and *ma'ii*, their name for Old Man Coyote, or between the principle of "disorder" Coyotism implied and the special coyote power present in all wild coyotes. So impossible was any of this to tease out that Navajos used the noun *ma'ii* to refer to all these qualities. Their word *ma'iitsoh*, meaning "large coyote," designated the wolf.

Farther north, on the Northern Plains of the mid-nineteenth century, both mountain man Osborne Russell and the Jesuit missionary Father Pierre-Jean De Smet wrote that the tribes there spoke of the coyote as Medicine Wolf, an animal helper able to "make things happen." Father De Smet, a Belgian Catholic who established numerous missions across the Northwest, was a close observer of Indian religions. He wrote in his memoirs that the northern tribes treated the coyote "as a sort of Manitou. They watch its yelpings during the night, and the superstitious conjurers pretend to understand and interpret them. According to the loudness, frequency, and other modifications of these yelpings, they interpret that either friends or foes approach."

One example of the Indians' sense of coyote power available to them through Coyotism famously occurred among the Navajos during the greatest misfortune that ever befell them. A tribe of

Pueblo Indian rock art of a coyote and its prey.
Courtesy Dan Flores.

hunters, herders, and raiders from the north who arrived some six hundred years ago in the Southwest's Four Corners, the ancient homeland of coyote evolution, the Navajos found themselves at war with US troops during most of the 1850s and early 1860s. Distracted by the Civil War, the United States, in a fit of exasperation at the success of Navajo raids, sent Taos mountain man and scout Kit Carson to command an invasion of Navajo country in 1863. Carson's men conducted a horrific scorched-earth campaign. By 1864 some 8,000 Navajos had surrendered to the frontier army, only to find themselves condemned to incarceration in eastern New Mexico, three hundred miles from home. Their "Long Walk" to the Bosque Redondo prison camp and four years imprisoned there under constant guard is a searingly painful chapter in Navajo history.

But Navajos also recall how this episode ended. After years of pleading to return home and frequent breakouts of small groups fleeing westward across New Mexico, in 1868 the United States finally agreed to a treaty that gave the Navajos a reservation and allowed them to return to their homeland. In Navajo oral tradition, the act that accomplished this longed-for release was not negotiation or pleading. It was their ritual performance of a Coyote Way ceremony, which infused Navajo leaders with enough "Coyote power" finally to effect their release.

Coyote power: surviving by one's intelligence and wits when others cannot; embracing existence in a mad, dancing, laughing, sympathetic expression of pure joy at evading the grimmest of fates; exulting in sheer aliveness; recognizing our shortcomings with rueful chagrin. These are the values Old Man America has embodied for thousands of years. Since his origins in the mists of continental human history all the way through the American counterculture of the 1950s, 1960s, and 1970s, Coyote has been an irresistible character. All creators of Coyote stories, from Paleolithic times until now, have instinctively grasped one essential: that humans delight in Coyote's exploits because we recognize how they shine a light on motives we struggle with. Coyote Man has always showed us truth.

Old Man America had been going along, having many a wonderful adventure and starring in his own biography of immortality for maybe 10,000 years. But two hundred years ago, his world began to change. Coyote's effortless trot across the continent had led him directly into the path of a whole new class of storytellers with names like Meriwether Lewis, Mark Twain, and Ernest Thompson Seton. While they sometimes found him as quirky and funny as ever, his divine glow of immortality, his role as humanity's holy antihero, began to dim. Not among the Indian people who had cherished him from those faint beginnings far back in the mists of the ancient continent, of course. But at least for a time, at the hands of these new inhabitants, Old Man America was about to get seriously demoted.

CHAPTER 2

Prairie Wolf

Once, for twenty years, I lived on the American Great Plains. I have never been so impressed with a landscape. If you can, suspend disbelief over that sentence for a moment. Like most Americans, I grew up in towns surrounded by forests; I, too, feel the undertow of a universal, hypnotic attraction to ocean beaches. I've also spent a fair share of my adult life embedded in the Rocky Mountains, a vertical base reversal from the horizontal world of the plains. But I've never gotten over the sense that the sea, the woods, the mountains all suffer in comparison with the prairie. Face-to-face, the vast prairie sweeps belie your instincts about such country. Their sublimity, I think, arises from their unfathomable boundaries and their self-confident grandness of scale, combined with an echoless, calm monotony of sensory affect.

But the reason the open steppes of the world so readily pull us up short probably has as much or more to do with genetic memory as anything else. For the past 45,000 years, since we left Africa and began to explore the planet, taking the measure of one landscape after another, we have been searching for a place that seems to have haunted our dreams. Almost certainly that dreamscape is our point of origin. In the literature of exploration, the landscapes that arouse our strongest passions always resemble our original African template: yellow savannahs speckled to the limits of our sight with herds and packs of wild animals. It is a phenomenon common to people across the globe.

So think of the American Great Plains as one more reminder that we can find home again. Two hundred years ago, except East Africa itself, no part of the globe thrilled us in the same primeval way. And it was out on these vast, horizontal, yellow sweeps, in the midst of grassland bison herds and packs of various wild canids, among flapping, hooting, scavenging birds, that the indigenous North American coyote first captivated western Indians with its intelligence and eerie familiarity. Now, in the nineteenth century, Americans from European and other backgrounds arriving at the edges of the great prairies were about to have their turn at understanding the coyote. This was where American and European scientific and literary travelers, rediscovering their home base, recorded their very first literary descriptions of the intriguing, jackal-like carnivore they at first called a "prairie wolf." In the process they would register an initial "contact" reaction to this unfamiliar predator. Their first-impressions take on the coyote would color the next chapters of its biography in bold, primary hues.

The months of August and September 1804 loom large in the natural history of North America, and indeed in the history of science worldwide. In the short stretch of three weeks, ascending the Missouri River into the grassy plains of today's Nebraska and South Dakota, American explorers Meriwether Lewis and William Clark described most of the characteristic wildlife that made North America unique in the world. Encouraged by President Thomas Jefferson to seek out and collect plants and animals not found in the Atlantic states, on August 23 the party downed the first bison most of them had ever seen. By September 7 they had seen their first prairie dogs, or "burrowing rats" in Clark's rather less flattering description. A week later they were marveling at a "Buck Goat of this countrey . . . more like the Antilope or Gazella of Africa than any other Species of Goat." That was the pronghorn antelope. Three days later they sighted "a curious kind of Deer of a Dark Gray Colr . . . the ears large & long." Thus did the American mule deer come to the notice of world science. The next day, somewhere in the vicinity of present Chamberlain, South Dakota, one more American original emerged from the wilds of the Great Plains. This one would eventually turn the heads of, even perplex, scientists in places as far-flung as Philadelphia, Paris, London, and Stockholm.

For most of September 1804, members of the Lewis and Clark expedition had reported seeing what they all assumed to be some kind of fox. The more they observed the sleek, beautiful canines, however, the less foxlike they seemed. So on September 17, a Monday on which the party's hunters killed thirteen white-tailed deer, two mule deer, three buffalo, and eight pronghorns, one of the hunters also shot an animal that a member of the group, Joseph Whitehouse, designated in his journal as a "Priari Wolf."

As a man of Enlightenment science, aware that his explorers' discoveries would be scrutinized in the great royal academies in Europe, Jefferson had encouraged precise scientific reporting. But the creature in the grass before him mystified William Clark. Although in size it was about like "a gray fox," viewed close up it did not appear very foxlike. Obviously it was not one of the large wolves these Americans knew from both Europe and the Eastern US. Agreeing with other members of the party who had gathered around to puzzle over the animal, Clark decided that it was a small carnivorous wolf entirely new to American science and resolved to name it the "prairie wolff." He went on to correct earlier recordings in his journal: "What has been taken heretofore for the Fox was those wolves, and no Foxes has been seen."

More than half a year later, on May 5, 1805, after the Americans had pushed across the plains as far as eastern Montana, Meriwether Lewis summed up the party's impressions of these new "prairie wolffs." His training in the sciences may have been cursory, but Lewis was careful and observant, and he knew this was the first bit of natural history about the new animal that much of the world would read.

> the small woolf or burrowing dog of the prairies are the inhabitants almost invariably of the open plains; they usually associate in bands of ten or twelve sometimes more and burrow near some pass or place much frequented by game; not being able alone to take a deer or goat they are rarely ever found alone but hunt in bands; they frequently watch and seize their prey near their burrows; in these burrows they raise their young and to them they also resort when pursued; when a person approaches them they frequently bark, their note being precisely that of the small dog. they are of an intermediate size between that of the fox and dog, very active fleet and delicately formed; the (years) ears large erect and pointed the head long and pointed more like that of the fox; tale long (and bushey); the hair and fur also resembles the fox tho' is much coarser and inferior. they are of a pale redish brown colour. the eye of a deep sea green colour small and piercing. their tallons are reather longer than those of the ordinary wolf or that common to the atlantic states.

Neither Lewis nor Clark ever wrote up a more detailed scientific description. Lewis rarely if ever proposed Latin binomials for their discoveries, which zoological taxonomy would have required to make them the official discoverers in Western science. Nonetheless, Clark's original 1804 description did offer up three tantalizing elements of the coyote's story that are still with us. First, this was an animal, unlike larger wolves and smaller foxes, unknown to Europeans, who thus did not bring centuries' worth of preloaded myths and stories about coyotes to North America, as they did with wolves. Second, Clark and his compatriots found the coyote confusing. Was it a fox? A wolf? Something else? Puzzled ambiguity has played a role in the coyote's biography ever since. Finally, and self-evidently, nineteenth-century Americans had to get halfway across the continent, to the edges of the Great Plains, before they began to encounter these prairie wolves. When the coyote first came to the attention of US citizens, then, this canine that no one expected was exclusively a creature of the western half of the country and, more specifically, of the Great Plains and the western deserts.

Neither Clark nor any other western traveler about to encounter coyotes for the first time would realize that the coyotes they came across weren't just one-way subjects of observation. Almost surely the explorers underestimated the animal, for coyotes were also taking their measure and gauging the possibilities. Contact with Americans was a propitious event from the coyote perspective as well.

As they so often turned out to be, Lewis and Clark were late to the game. Other Europeans were studying the natural history of the Americas too, and they had been at the task far longer. So the first printed description of a coyote appeared not in the 1814 publication of Lewis and Clark's journals but in a work that saw print some 160 years earlier. In 1651 Spanish author Francisco Hernandez, in a book chapter titled "Concerning the Coyotl, or Indian Fox," was actually the first Western author to introduce the North American coyote to a reading audience.

In the coyote's initial audition as a literary figure for Europeans, confusion was naturally the theme. This "is an animal unknown to the Old World," Hernandez wrote, and it was either a fox, an "Adipus" (a Mediterranean term for a jackal), or some new, distinct species. He continued that it had a "wolflike head" but "approaches in appearance our own fox," and "its bite is harmful." This first written notice of coyotes by a European spoke of the animal with some alarm, pointing out that it not only preyed on the sheep Europeans had stocked in New Spain but also attacked "stags and sometimes even men." Hernandez concluded, "The animal inhabits many regions in New Spain, particularly those tending toward cold and chill climate." He also included these curious lines: the new creature, he said, was "a persevering revenger of injuries" but also "grateful to those who do well by it." He used the ancient Aztec name for it: *coyotl*.

Subsequent Spanish authorities, such as Francisco Javier Clavijero, in his *Historia Antigua de Mejico*, published in 1780—a quarter century before Lewis and Clark named the animal "prairie wolf"—still relied on Hernandez. But by Clavijero's time the Spaniards had more experience with American animals, so his account expands the coyote's treatment considerably: "The *coyotl*, or *coyote*, as the Spanish call it, is an animal similar to the wolf in its voracity, to the fox in its cunning, to the dog in its shape, and in other propensities to the *adive*, or *jackal*; for which reason some Mexican writers have counted it among several of these species; but it is undoubtedly different from all of them."

Clavijero also noted that by 1780 the *coyotl* had become "one of the most common quadrupeds of Mexico."

Indians may have had thousands of years' worth of close familiarity with the coyote, and prior accounts and descriptions by Spanish authors might exist, but the taxonomic approach that dominated Western science by the early nineteenth century was clear. To claim official scientific discovery of a species, one had to publish a detailed description accompanied by a proffered binomial to classify it in the Linnaean system. As a result, the honor of adding the prairie wolf to science fell not to William Clark or even Francisco Hernandez but to a naturalist named Thomas Say, the zoologist on an American exploring expedition to the West some fifteen years after Lewis and Clark. Say, a young naturalist from the Academy of Natural Sciences of Philadelphia, was a member of the Stephen Long expedition across the plains to the Rocky Mountains of Colorado in 1819 and 1820. Not until 1823, however, did he officially describe the "type specimen" of the prairie wolf from a Nebraska coyote he was finally able to catch by using a bobcat as bait. His binomial, *Canis latrans*

Say (Say's barking canine), has been the recognized scientific name ever since, although not—it turned out—without challenge.

Because Say became the "official" discoverer of this "new" carnivorous canine, his journal entries are worth mulling over for insights into how this strange animal first struck Americans. "The prairie wolves roam over the plains in considerable numbers," he began his account. Not only were they "by far the most numerous of our wolves," but they constantly loitered around the explorers' camps, seemingly very curious and unafraid and affording western travelers many opportunities to study their habits.

Say became the first writer to notice and comment on what we today recognize as coyote fission-fusion adaptations. He wrote that although they appeared in singles and pairs, coyotes "often unite in packs for the purpose of chasing deer, which they very frequently succeed in running down, and killing." Their "swiftness and cunning" were often insufficient to bring down large prey, however, so they were "sometimes reduced to the necessity of eating wild plums, and other fruits . . . in order to distend the stomach, and appease in a degree the cravings of hunger."

The naturalist could not help but notice, as had Meriwether Lewis, a defining prairie wolf trait: "Their bark is much more distinctly like that of the domestic dog, than of any other animal; in fact the first two or three notes could not be distinguished from the bark of a small terrier, but these notes are succeeded by a lengthened scream." To that he added, in an initial rendering of what was about to become a constant refrain about coyotes, "The wonderful intelligence of this animal, is well worthy of note."

As for the prairie wolf's place among the world's carnivorous canines, Say was obliged to consider the widespread claim made by European naturalists, on reading the Lewis and Clark descriptions, that the prairie wolf was most likely an American species of jackal. This became a serious and lengthy debate. Defending his conviction that this creature was wholly unknown to Western science, Say noted, "The *latrans* does not diffuse the offensive odour, so remarkable in the two species of jackalls, (*C. aureus* and *C. anthus,*) though in many respects it resembles those animals." That observation turned out to be insufficient to lay to rest arguments that the American prairie wolf might actually be a new form of jackal.

Say ended his description with an intriguing claim: "This animal . . . is most probably the original of the domestic dog, so common in the villages of the Indians of this region, some of the varieties of which, still retain much of the habit, and manners of this species."

In combination with expedition artist Titian Ramsay Peale, Thomas Say briefly planned a book that would have been an early mammal version of John James Audubon's *Birds of America*, combining the naturalist's descriptions with Peale's paintings. The book never materialized, but it does seem remarkable in light of their plans for it that Peale's 1819 watercolor painting of the prairie wolf—the very first visual rendition of a coyote by an American artist—is titled "Fox." At least Peale didn't call it a jackal.

Today Thomas Say receives unquestioned credit for adding the coyote to Linnaean science, but given the number of naturalists drawn to the West in the early nineteenth century, he not surprisingly had rivals in the game. Close, almost unaccountable misses to describe the animal provide additional details about the coyote's original circumstances. One of these involved one of the most famous English naturalists ever to work in early America, Thomas Nuttall, whose coyote misses are especially intriguing.

At a time when American universities were only barely starting to turn out trained field naturalists (that's why no bona fide naturalist accompanied Lewis and Clark), Nuttall arrived from Yorkshire in 1808 at the age of twenty-two and at once came to the attention of a prominent American professor of the natural sciences, Benjamin Smith Barton of the University of Pennsylvania. Barton had been Jefferson's advisor on natural sciences for the president's western expeditions; one of his students, Dr. Peter Custis, had accompanied the president's 1806 expedition to the edge of the Southwest. Barton became a Nuttall advocate, introducing him to William Maclure of the American Philosophical Society, who nominated him for

The first painting of a coyote by an American naturalist,
Titian Ramsay Peale, misidentified the animal as a fox!
Courtesy American Philosophical Society.

membership. With patrons like these, Nuttall got to explore across the West for the next three decades.

In 1818 and 1819 Nuttall embarked on a natural history expedition to the edge of the southern prairies. It was the same summer that Thomas Say first saw coyotes a few hundred miles farther north. Nuttall's route took him diagonally (southeast to northwest) across Arkansas Territory, then westward as far as central Oklahoma, and on south to the Red River and the border of Spanish Texas.

Read Nuttall's southwestern journal today, and it's clear he was naturalizing along the logical eastern edge of the coyote's early nineteenth-century range. Everywhere these early explorers first encountered them, coyotes were part of a zoological cohort that included bison, pronghorns, prairie dogs, and usually elk. As travelers moved from the east and began to enter the West, the presence of all these animals coincided with the first appearances of a grassland-dominated landscape. Nineteenth-century Americans were fascinated sensory observers of these changes and repeatedly recorded a predictable progression. Explorer John C. Fremont's journey westward from Chouteau's Landing on the Missouri in 1842 captures it well. A month's travel up the Kansas River, as the woodlands opened into sweeping, grassy prairies, Fremont began to report pronghorns "running over the hills." The next day he noticed that artemesia, or sagebrush, had become common. The day after that, elk began to appear on the river. Within a week they were among the buffalo herds, and at that point "wolves in great numbers surrounded us during the night . . . howling and trotting about."

Almost every traveler who entered the Great Plains from the east had this experience, and coyotes—"small wolves"—always attracted attention in the midst of this zoological sea change. "Three different kinds were present," Francis Parkman wrote. "There were the white wolves and the gray wolves, both extremely large, and besides these the small prairie wolves, not much bigger than spaniels." Parkman saw them in action: "They would howl and fight in a crowd around a single carcass, yet they were so watchful and their senses so acute, that I was never able to crawl within a fair shooting distance; whenever I attempted it they would all scatter at once and glide silently away through the tall grass."

The ecological edge that marked this profound environmental boundary sometimes transitioned the country from "the East" to "the West" in no more than twenty or thirty miles.

So it fascinates me that in 1819 Thomas Nuttall traveled right down the seam of that East/West boundary, from the present Tulsa area southward along the western stretches of the Ouachita Mountains to the Red River, moving in and out of prairie country like "an immense meadow . . . covered with luxuriant herbage, and beautifully decorated with flowers." Infrequently the Englishman saw bison, the most common mammal marker of the prairie edge. But he never mentions coyotes.

In August 1819, in company with a plainsman named Lee, Nuttall decided to explore farther westward. On the Cimarron River that month, just northeast of present-day Oklahoma City, the naturalist scribbled in his journal that they had seen a badger den "about the size of those made by the prairie wolf." So by this point he had seen coyote dens. But Nuttall had taken seriously ill, fainting and nearly falling from his horse in a malarial delirium. This was his only reference to an animal that in 1819 was his to describe and win credit for, had he only acted.

Sixteen years later, having traveled across the West with fellow naturalist John Kirk Townsend, on California's Central Coast near Monterey, Nuttall listened as coyotes "tame as dogs yelled every night through the villages." By that point a fellow naturalist exploring the exact coyote prairie edge that Nuttall visited in 1819 had set off a scientific debate about the strange new canines of the prairies.

Was the new animal actually a jackal, an Old World creature known until that time only in Africa, the Middle East, and southeastern

Prairie Wolf, Fort Clark, 1833–1834, by Karl Bodmer.
Courtesy Joslyn Art Museum.

Europe? That was the first question science posed about "prairie wolves," and from the 1830s to the 1850s, the debate set the naturalists of the Philadelphia Academy against those of the new Smithsonian in the capital city. Earlier in western exploration, the academy had supplied many of the naturalists. Through the influence of a scientist named Spencer Baird, however, the Smithsonian ended up manning the federal government's Mexican-Boundary Surveys and Pacific Railroad Surveys with botanists, mammalogists, ornithologists, and geologists of its choosing.

Initially, Say's designation of the prairie wolf as a new species, previously unknown and unrelated to jackals, carried the day. And Say had an early and influential supporter, European explorer Prince Maximilian of Wied-Neuwied, who, it so happened, was about to make a famous journey up the Missouri River in the 1830s, right into the heart of prairie wolf country.

The early-nineteenth-century American West attracted European naturalists and artists in much the way East Africa did later in the century and for many of the same reasons. European scientists, trained in Linnaean taxonomy and steeped in the romantic hero-adventurer tradition of Prussian naturalist Alexander von Humboldt, looked to the American West and to the plains of East Africa as one final chance for the holy experience of the wild Pleistocene planet. Landlocked Germans, denied sea exploration, were especially avid consumers of the reports of these terrestrial adventures in wild places.

The Carl Sagan or Neil deGrasse Tyson of his day, Alexander von Humboldt was example, model, and mentor to many of the explorers from Europe; some of them, it turns out, were apparently also his lovers. Among the young men who won von Humboldt's personal patronage, Prince Maximilian of Wied-Neuwied, who explored Brazil from 1815 to 1817 and then ascended the Missouri River into the American West in 1833 and 1834, became one of the most celebrated of his protégés. Maximilian had received anthropological training at the University of Göttingen. As a result native peoples were always his real interest. But nineteenth-century explorers had to be generalists, so he taught himself the basics of natural history as well.

Prince Maximilian also had the good fortune to spend the 1832–1833 winter in America's most personable science and arts commune, New Harmony in Indiana. Among his confidantes on all-things West that winter was Thomas Say, the Philadelphia Academy naturalist whose *Canis latrans* designation had established the prairie wolf as a new species and no jackal. You would guess that the German prince heard a good deal about Say's barking canid of the plains that winter in Indiana.

Maximilian also had the good instincts to retain the services of the best-trained painter to venture into the West until that time. Young Karl Bodmer was a Swiss artist whose voluminous work on the prince's Missouri River voyage would become one of the best of all visual representations of the natural West. Urged by his patron to pay close attention to ethnographic detail, Bodmer produced Indian portraits and scenes of plains life so exacting that both modern Indians and Hollywood have relied extensively on them to recapture the world of the Plains Indian. He also brought a skill to capturing the atmospherics and exposed geology of western landscapes that went unmatched until Maynard Dixon traveled to the West seventy-five years later. But Bodmer's real training was in portraying the wildlife of the European forests. So it is no accident that this young Swiss artist drew the best portrait of a coyote produced by anyone in the early West.

By the late summer of 1833, Maximilian and Bodmer were at Fort McKenzie, in present-day eastern Montana, where, as the prince wrote, "we everywhere met with wolves." As an acquaintance of Say's now and a scientist who kept up in his field, Maximilian properly understood the difference between gray wolves and coyotes. "In the environs of the fort there were, at this time, wolves, foxes, and a few hares," he wrote, "and during the night we heard the barking of the prairie wolves (*Canis latrans* Say), which prowled about, looking for any remnants of provisions." One of these, trapped for the sake of art and science, became the model for Bodmer's subsequent gorgeous pencil portrait, a gracefully and precisely rendered representation on paper of a western coyote's sense of self.

Prince Maximilian went on to write a classic of the early West with his two-volume *Travels in the Interior of North America*, which appeared in print in Germany between 1839 and 1841. For his part, Bodmer became one of the most celebrated wildlife and forest-scene artists of nineteenth-century Europe. His "Prairie Wolf" from 1833 North America, done before American culture would spin out a much more negative interpretation of the coyote, stands today as one of the most exacting wild canid portraits ever made.

Maximilian's (and by extension von Humboldt's) acceptance of Say's argument—that the prairie wolf was a wholly new animal, unknown in world science—did not settle the jackal issue. Following the Mexican War and the Treaty of Guadalupe-Hidalgo, which added a third of the present United States to the nation, American scientific explorations to the West intensified. The naturalists who accompanied the expeditions tended to come out of Smithsonian connections, although some Philadelphia Academy men were still involved. One of the latter was a physician-naturalist named Samuel Washington Woodhouse, who worked on three western expeditions in the 1850s, including a famous early exploration of the Grand Canyon with Captain Lorenzo Sitgreaves But prior to that expedition, Woodhouse had explored with Lieutenant Israel Woodruff in the Indian Territory—as it turned out, in exactly the same country where Thomas Nuttall had missed joining the jackal debate back in 1819.

During Dr. Woodhouse's naturalizing with the Corps of Topographical Engineers in Oklahoma, he had a coyote encounter, and it was enough to draw him into the debate against his fellow naturalist, Thomas Say. Just twenty-seven and a native Philadelphian, Woodhouse obviously knew about Say's *Canis latrans*. He probably knew that Thomas Nuttall had preceded him on the Cimarron River, where Woodhouse began to collect in October 1849 as the party entered the prairie country. They saw their first signs of buffalo just east of present-day Stillwater, near a landmark called Bald Eagle Mound

Richard Kern's drawing of naturalist
Samuel Woodhouse's North American Jackal.
Courtesy US Government Printing Office's "Sitgreaves Report."

and where, on October 19, they terminated the season's work and returned to their homes for the winter.

In the summer of 1850, the party resumed the survey at Bald Eagle Mound. Camped now in that magical edge-of-the-plains boundary, Woodhouse got his opportunity to join the debate in July. In that same week he encountered elk, a wolf, and his first bison. On July 17, two of his party brought in "2 prarie wolves" they had shot, too late in the day to preserve. Maybe Woodhouse thought all night about his decision, but the next day, as he prepared the specimens, he concluded that he would challenge Say directly. With the coyotes in hand for close examination, he agreed that this was a species new to science, clearly in the genus *Canis*. But as far as Woodhouse could determine from the jackal descriptions in his texts, the North American canines he was handling were jackals. He decided to make the larger male his type specimen and named it *Canis frustror*, the "North American Jackal." Later Richard Kern, artist for several exploring expeditions into the Southwest, did a drawing of the animal for the official "Sitgreaves Report" of 1853, by which the "North American Jackal" entered science via the massive US commitment to publishing exploration accounts.

Woodhouse's challenge did not survive the test of science. Say prevailed, and zoology recognized the American prairie wolf as a new species that occupied a jackal-like niche on the Western Plains. Today we know that similar golden jackals and coyotes come from the same line of canid evolution in ancient America and went their separate ways only 1 million years ago. Taxonomists now recognize Woodhouse's "jackals" as a coyote subspecies, *Canis latrans frustror*. His "Holotype, specimen no. 4105" in the National Museum of Natural History shows a young male animal with an especially dark pelage—its tail is almost black. But it is clearly a coyote.

Jackal or wolf? This wasn't the only question about the prairie wolf that confronted the first generations of Americans who encountered the animal. They faced two more big questions, the second of them as fundamental as whether to classify the new animals as jackals. It revolved around what to call them.

In 1832 and 1833, as Prince Maximilian and Karl Bodmer were wintering at the New Harmony colony, farther south in the prairie wolf's range—Santa Fe and the Southern High Plains to be exact—poet and future judge Albert Pike of Boston was preparing to venture from New Mexico eastward into Comanche country. His forays resulted in an obscure but important little book, *Narratives of Two Journeys in the Prairie*, which has intrigued readers for nearly two hundred years now.

Pike's book came about in this way. A decade had passed since Mexico's successful revolution against Spain had opened the fabled cities of Santa Fe and Taos to American traders and mountain men like Kit Carson. By the time Pike got there, a dozen years of trapper work in the Southern Rockies had made beavers scarce in the mountains, so the literary-minded New Englander had joined a group of forty-five trappers (including famous preacher-trapper Old Bill Williams) heading east in what proved a vain and foolhardy search for beaver colonies out on the High Plains.

The importance of Pike's little book for coyote history, though, has less to do with beaver or adventure than with the discovery of an ancient name out of old America. Traveling in a party equally divided between Americans and New Mexicans, a few weeks onto the prairies, Pike bent over his journal one night to pen a description of the wild canids the party had seen, and for the first time an English-speaking writer gave the world access to the old continental name for the animal Americans had been calling a prairie wolf. On these prairies, Pike reported, were "bands of white, snow-like wolves prowling about, accompanied by the little gray *collotes* or prairie wolves, who are as rapacious and as noisy as their bigger brethren."

One encounters the word "collotes" nowhere else, and its pronunciation must have confused Pike's readers. I think the truth was something like this: as a resident of New Mexico for a year, traveling with New Mexicans, and a student of languages himself, Pike was not offering a phonetic rendering. Instead he was using the Spanish double *l*, which is pronounced as a *y*.

This was the first use of a form of the word "coyote" in printed English. But the book that put the name into its modern spelling and more widely alerted Americans and Europeans to the fact that the prairie wolf had another, even more evocative name was Josiah Gregg's classic *Commerce of the Prairies*, first published in 1844.

Gregg was a Missourian who had followed two brothers into the Santa Fe trade and eclipsed both (and almost everyone else involved) to become a trader, amateur naturalist, and author. Gregg's life well illustrates the freewheeling society of the American frontier. He studied, successively, surveying, medicine, and the law before "consumption" (tuberculosis) led a physician to prescribe a climate cure—a trip to the Southwest—for the then twenty-four-year-old. Mexico had opened up storied Santa Fe to trade with the United States in 1821, so the easiest way to get to a dry climate was to join a trading party to the fabled southwestern city.

Gregg made his first journey across the plains in 1831 and for the next nine years crossed the western prairies repeatedly, religiously taking notes on everything he saw and heard, with the idea of writing a book on the region. The result was his *Commerce of the Prairies*, a famous, essential firsthand account of the nineteenth-century West, which went through fourteen printings in its first half century and was published in both England and Germany.

A good number of Gregg's associates thought he was a crank, probably a hypochondriac, and for certain a pseudointellectual. He may have been all of those things, but he was also a careful observer of nature who later corresponded with and collected plants for famed St. Louis botanist George Engelmann. In the late 1840s, Gregg collected more than 650 plants in Mexico; today twenty-three plant species are named for him. So his observations of the animal he said the American traders called the prairie wolf are worth a close look, not only for his early natural history of the animal on the wild plains but for the insights his account yields into how the name "prairie wolf" over time gave way to "coyote."

Gregg and other American traders entered a New Mexico where Spaniards had settled among the Pueblo Indians along the Rio Grande more than two centuries earlier. The Spanish *entradas* that founded Santa Fe in 1610, however, were actually multicultural. Sephardic Jews, fleeing the Inquisition to the far ends of the Spanish Empire, were among the settlers. So were Mexican Indian auxiliaries, fighting under the command of Spanish officers from Barcelona and Madrid. Some of those Indian New Mexicans spoke Nahuatl, the ancient language spoken by many of the Aztec Empire's subjects. These people had a timeless familiarity with the animal the Americans were calling a prairie wolf. The Nahuatl name, 1,000 years old by the 1830s, was of course *coyotl*.

Gregg and Pike put us in the place and time when Americans discovered the coyote's old name. Gregg bartered and traveled with New Mexicans and had to have encountered prairie wolves in their presence. Both Americans would have heard the translation of the Nahuatl singular noun into Spanish. Hispanicizing the Aztec word *coyotl* was accomplished by dropping Nahuatl's absolutive suffix *-tl* and substituting a Latin-accented *e*. So the new name Gregg and other American traders heard applied to the animals by the residents of Santa Fe was *coyote*, pronounced by nineteenth-century Spanish speakers as *coy-YOH-tay*, with the accent on the second syllable.

Gregg introduced the new name for the prairie wolf to a global audience almost matter-of-factly, as a lead-in to the natural history he obviously found far more intriguing than the etymology of the name: "There is a small species [of wolf] called the *prairie wolf* on the frontier, and *coyote* by the Mexicans, which is also found in immense numbers on the Plains. It is rather smaller than an ordinary dog, nearly the color of the common gray wolf, and though as rapacious as the larger kind, it seems too cowardly to attack stout game. It therefore lives upon the remains of buffalo killed by hunters and by the large wolves, added to such small game as hares, prairie dogs, etc., and even reptiles and insects."

Gregg was enough a student of natural history to know about the ongoing scientific debate: "The coyote has been denominated the 'jackal of the Prairies;' indeed, some have reckoned it really a species of that animal, yet it would seem improperly, as this creature partakes much less of the nature of the jackal than of the common wolf." He continued with an observation that everyone who lives in Coyote America can relate to: "Like ventriloquists, a pair of these will represent a dozen distinct voices in such succession—will bark, chatter, yelp, whine, and howl in such variety of note, that one would fancy a score of them at hand. This, added to the long and doleful bugle-note of the large wolf, which often accompanies it, sometimes makes a night upon the Prairies perfectly hideous."

Finally, Gregg weighed in with a prairie legend that flirted with a larger truth than he ever imagined: "Some hunters assert that the coyote and the dog will breed together. Be this as it may, certain it is that the Indian dogs have a wonderfully wolfish appearance."

Pike's and Gregg's books were the first to teach Americans the old continental name for the prairie wolf. Being willing to attempt in everyday frontier English a multisyllabic Nahuatl noun that had itself been Hispanicized was another matter. Help with unraveling how that oral-learning process worked comes from another Southwestern writer of the time, George Frederick Ruxton. A minor English nobleman and a military veteran at the age of seventeen, Ruxton had carved out an arc of adventures from Spain and Africa to Canada before journeying, in 1846, to Santa Fe, where for the next three years he traveled with a diverse crowd of traders and mountain men operating out of northern New Mexico. He spent much of his time in present-day Colorado, particularly in the high mountain valley the trappers called Bayou Salado, now South Park. The Englishman was self-aware enough to know that some back home thought he'd reverted to barbarism. But for Ruxton, "the very happiest moments of my life have been spent in the wilderness of the Far West."

Ruxton's *Life in the Far West* is a strange creation, a form of novelized history that folds together both his own experiences and those of the reckless hunters he consorted with. He had enough experiences himself that the affection his book evinces for coyotes seems genuine. Because Ruxton entered the West where he did, like Pike and Gregg, he learned to call prairie wolves by their Aztec name. Like Gregg, whose book he had no doubt read, Ruxton spelled it "coyote." But that was only one of his spellings of the word. Another he used just as frequently offers up hints about how English- and French-speaking Americans, very early on, put their own stamp on the animal's name.

As *Life in the Far West* proceeds, Ruxton increasingly spells "coyote" slightly differently, finally settling on a spelling that I think more accurately reflects what he was hearing on the trail and around the campfires among his associates: "cayeute" was the spelling he came to prefer. This second version, I think, phonetically renders how English speakers were saying the word. If we can trust Ruxton's ear—after all, the Englishman famously took great pains to capture the vernacular language of the trappers—by the 1840s Americans on the frontier were already anglicizing "coyote." Or more accurately, they were anglicizing a Hispanicized Indian word. In their parlance *coy-yo-tay* was evolving into a word that to Ruxton's ears sounded like *KI-oht*—two syllables, with the accent on the first, Anglo-Saxon-like.

Here is what Ruxton wrote about his cayeutes: "Besides the buffalo wolf, there are four distinct varieties common to the plains, and all more or less attendant upon the buffalo. . . . [L]ast and least [is] the *coyote*, or *cayeute* of the mountaineers, the 'wach-unkamnet,' or the 'medicine wolf' of the Indians, who hold the latter animal in reverential awe." No jackal, this, Ruxton said, was a "wolf, whose fur is of great thickness and beauty." It was "of diminutive size," to be sure, but was "wonderfully sagacious, and makes up by cunning what it wants in physical strength." Smart as it was, "the cayeute, however, is often made a tool of by his larger brethren, unless, indeed, he acts from motives of spontaneous charity." He meant that since coyotes were less anxious around humans than gray wolves, hunters often threw them scraps. But then "the large wolf pounces with a growl upon him, and the cayeute, dropping the meat, returns to his former position, and will continue his charitable act as long as the hunter pleases to supply him."

Americans and Europeans learned from *Commerce of the Prairies* and *Life in the Far West* that Lewis and Clark's prairie wolf actually possessed a very old and rather exotic name, one that much predated the European arrival in America. So from the late 1840s on, "coyote" gradually obliterated "prairie wolf" as the proper name for the still mostly unfamiliar new wild dog of America. Herman Melville's line in *Moby-Dick* invoking "the bloodshot eyes of the prairie wolves" was one of the final appearances of Lewis and Clark's chosen name in print. That was in 1851.

The third question nineteenth-century Americans posed about the prairie-wolf-become-coyote was as fundamental as what to call it. What were citizens of the United States supposed to make of this unfamiliar little canine, about which they had no legends, no folklore, no preconceived notions? However "diminutive," as a carnivorous predator the coyote was arguably starting from a position of disadvantage. Was it, like the larger gray wolf, a menace, a threat to civilization, vermin to be eradicated? Might the coyote have useful, even commercial, qualities? Did it serve some function? Resolving those questions—determining the essential character of the coyote—would take the entire second half of the nineteenth century and most of the twentieth. We're still trying to figure out what we think about them even now.

At mid-century, there was no emerging consensus. Various travelers to the West observed coyotes, often with fascination, and sometimes painted them, but they initially took no stand on the animal's essential character.

When John James Audubon finally went west with his sons in 1843 to work on what he planned as his second great book on Amer-

Prairie Wolves, 1843, by John James Audubon.
Courtesy University of Michigan Special Collections Library.

ican nature, *The Viviparous Quadrupeds of North America*, he finally got to observe and paint what was then still called the prairie wolf. His oil painting of two specimens of *Canis latrans* (his endorsement of Say) is among the more successful mammal paintings to come from the Audubons' journey up the Missouri River. Audubon and his sons painted almost the full suite of Great Plains species on this trip, not just coyotes but bison, lions, prairie dogs, and even the Swift fox, which they realized was actually a kind of fox and not another version of the "prairie wolf."

The famous bird painter dutifully entered various observations about the coyotes he saw, but they were value-free: "We saw coming from the banks of the river no less than eighteen Wolves, which altogether did not cover a space of more than three or four yards, they were so crowded. Among them were two Prairie Wolves." A French hunter also told Audubon he had seen wild horses kill wolves—most likely coyotes—by seizing them in the middle of their backs and throwing them high enough to stun them, after which "they stamp[ed] upon their bodies with the fore feet until quite dead."

In the decades before a stereotype emerged, when Americans were seeing coyotes with fresh eyes, coyote observations were primarily curious and general. Like many predators, the animal had exotic qualities that drew attention to it, but aspersions, let alone hatred, seemed wasted on the little canines. Prairie wolves the size of "spaniels" hardly seemed worthy of special loathing to Francis Parkman, just annoyance that they were so wary before his gun. Washington Irving found that their howls "gave a dreariness to the surrounding solitude," but he didn't bother with further castigations. Heinrich Balduin Mollhausen, another German who trained in natural history as an acolyte and favorite of the great von Humboldt and served as a naturalist and artist on several American exploring expeditions in the 1850s (before going on to a career as "the German James Fenimore Cooper"), at one point in his adventures saved himself from starvation by subsisting on frozen wolf or coyote meat. One could argue that New Englander Herman Melville was still using "prairie wolf" as the preferred name because of his distance from coyote country. But at least he did get them into his book without casting aspersions.

When America's interest in settling the West revived after the Civil War, and explorers, government officials of various stripes, military troops, and literary adventurers—as well as immigrants wishing to settle—again moved into coyote country, a peculiar fascination set in with the animal most were seeing for the first time. On the trails, in churches and bars, and in the new farmsteads and ranch houses, coyotes came up for much discussion. Scottish naturalist Hans Kruuk argues that with our evolutionary background as hunters, we humans look on predators with an especial fascination as competition. Our evolutionary history holds a genetic memory of when we were prey too, so we can also exhibit an instinctive anticarnivore loathing. Because they seemed like smaller wolves, coyotes aroused suspicion in frontier folk. Though too small to arouse a prey response, they did strike us as potential competitors. That became an acutely realized reaction when American farmers and ranchers actually began to live and raise domestic livestock in coyote country.

With no imported mythology about them and scant or no interest in Indian religions or fables about a Coyote deity, Americans in the West found coyotes ripe for original interpretation. Beginning in the 1870s and lasting for the rest of the nineteenth century, a new, unflattering image soon formed in the American mind, and particularly in a very specific mind—that of writer-humorist Mark Twain.

Twain's description in his 1872 best seller *Roughing It* provided the foundation for a coyote assessment that began as neutral but started to grow worse as time went on. Although he was traveling by train, the author's first sighting of "the cayote, pronounced ky-o-te," he tells a reading public, was somewhere west of Omaha, in "vast expanses of level greensward, bright sunlight, an impressive solitude," on the same day he saw his first prairie dogs and pronghorns. But Twain's prose ignored both the latter in favor of a three-page

Prairie Wolf 77

Western coyote.
Courtesy Dan Flores.

soliloquy, verging on comic rant, about an animal—"not a pretty creature or respectable, either"—he came to know as he crossed the West.

In Twain's view, the coyote's choice of homes defined him, for he lived "chiefly in the most desolate and forbidding deserts." With the animal's suspect habitat laid out, Twain then hit stride in what became a classic description with a long reach in American culture: "The cayote is a long, slim, sick and sorry-looking skeleton, with a gray wolf-skin stretched over it, a tolerable bushy tail that forever sags down with a despairing expression of foresakenness and misery, a furtive and evil eye, and a long, sharp face, with slightly lifted lip and exposed teeth. He has a general slinking expression all over. The cayote is a living, breathing allegory of Want."

As was Twain's wont, he was just warming up. "The meanest creatures despise him, and even the fleas would desert him for a velocipede." Twain's coyote is "spiritless and cowardly," and "he is so homely!—so scrawny, and ribby, and coarse-haired, and pitiful." How could such a vermin subsist? As far as Twain could tell, he was a pure butcher house scavenger, living on offal, carrion, and the carcasses of emigrant train livestock.

But the coyote had a sense of humor, and what other writer would have been more appreciative of that? Send a dog that has a good opinion of itself after a coyote, Twain averred, and the "cayote will go swinging gently off on that deceitful trot of his, and every little while he will smile a fraudful smile over his shoulder that will fill that dog entirely full of encouragement and worldly ambition." Jokester coyote would gleefully let the town dog get within twenty, even six feet of him, but despite the tumult at his heels, "the cayote glides along and never pants or sweats or ceases to smile. . . . [A]nd forthwith there is a rushing sound, and the sudden splitting of a long crack through the atmosphere, and behold that dog is solitary and alone in the midst of a vast solitude."

Gone now was the Indian deity who created North America. Gone was even the perplexing prairie wolf of earlier in the century. Observing the same animal as before, Americans now saw a sick, despairing, forsaken, miserable creature, one (as we ourselves warmed to the task) that took on the unsavory traits of cowardice, cunning, and cruelty. In the late nineteenth and early twentieth centuries, a repetitive trope emerged. In New York journalist Horace Greeley's version, the coyote was "a sneaking, cowardly little wretch." By the time English adventuress-writer Isabella Bird wrote *A Lady's Life in the Rocky Mountains* in 1879, that description was something of a set piece: "I saw two prairie wolves, like jackals, with gray fur, cowardly creatures, which fled from me with long leaps."

Within a decade, articles like Ernest Ingersoll's "The Hound of the Plains" in *Popular Science Monthly* (1887), and later Edwin Sabin's "The Coyote" in *Overland Monthly* (1908), were describing coyotes as "contemptible" and "especially perverse." Their howls were "eerie" and "blood-stilling," even defiant. Coyotes supposedly lacked "higher morals," and naturally they were "cowardly to the last degree." Exploring ideas for commercial gain from the killing of coyotes, a 1920 article in *Scientific American* asserted that coyotes were not worth the price of the ammunition to shoot them, then went on to add the ultimate insult for the age (and a patriotic reason for shooting them anyway): the coyote, the writer avowed, was the "ORIGINAL BOLSHEVIK."

From avatar Coyote deity to avatar Original Bolshevik, the coyote had traveled a long way in just a century's time. Wherever his destination lay, the road was now, suddenly, a very steep uphill climb.

CHAPTER 3

A War on Wild Things

Living in the evolutionary heartland of America's native canines, as I have for a decade in the piñon-juniper mesas south of Santa Fe, I have borne witness to one certain truth about coyotes as neighbors: you do not see them so much as hear them. Even in rural New Mexico I only see a coyote trot through the yard or lope across the road in front of my car perhaps once a month. But howling coyotes mark my nights almost without fail. I hear their salutations through my open windows or skylights often enough to awaken from my summer dreams. Yodeling coyote music is inseparable from the silvery wash of planets and the high moons of the winter night skies of this part of the world.

Sometimes one coyote's melody becomes a general regional symphony, as individuals and pairs and packs join in, and when that happens you can hear—or perhaps you imagine in the mind's eye and ear—coyote song spreading like a contagion, picked up by pack after pack until it fades into far distances, faint howls winking out in the mind in a kind of aural canine redshift. It is sometimes easy to think, in the summer New Mexico dusk, that what begins with a single coyote pouring out his soul across these canyons and dwarf forests

Young coyote howling North America's original national anthem.
Courtesy Dan Flores.

has by morning rippled in concentric circles from this spot of origin across the full sweep of North America, as far as the last coyote at the farthest edge of the last ripple. Which of course today could be in Maine. Or Alaska or Costa Rica or Florida.

In my view the coyote's howl is the original national anthem of North America, a canine "Star-Spangled Banner" that has been playing for nightly sporting events on this continent for nearly 1 million years.

But over the past couple of centuries, most Americans, at least judging by the literary types who wrote about the subject, formed quite different mental images when they heard a coyote howl. I won't belabor the small variations among the examples (and there are a very great many examples) because the general thrust is always the same. The coyote's cry, for many Americans riveted by the sound, did not intone the ballad of the continent—America's ancient native song—that some of us hear today. Instead, someone like Lieutenant Robert Carter, a New Englander and West Pointer who in 1935 wrote a book about his youth on the Great Plains, when "that section was overrun with Indians, buffalo, wolves, jack rabbits, prairie dogs, sage brush and cactus," viewed the howl of this indigenous canine as positively alien. Carter's description is so typical, it's a frontier cliché. "Their blood-curdling howls—which is first a sharp bark, followed by a succession of sharp, staccato yelps running into each other and ending in a sort of long drawn out quavering howl—were, at times almost indescribably melancholy, and awakened us at all hours of the night, which caused [my] newly-made bride to sit up in bed and shrink back in alarm."

A sense that so much about North America was strange and frightening and that we ought to terraform and remake it extended to every element of continental ecology, from grasses to animals of all kinds. But in truth, almost no other creature reaped the whirlwind of condescension and hostility toward "alien" American nature in quite the way coyotes did. We campaigned to erase those "manic, lunatic" howls for all time and good riddance. And even as evidence mounted of the wrongheadedness and futility of that course, we spent more than half a century in furious pursuit of it.

―――――

On September 8, 1887, the Salt Lake *Weekly Tribune* ran a piece with an unanticipated storyline. Rather than indulging in the usual coyote character assassination of the 1880s, the writer took a surprising tack. The coyote, he argued, "is the salvation of the Utah farmers in some sections. About Fillmore, a few years ago, in a co-operative way, they exterminated these wild canines by poison, since which time they are under the necessity of fencing their crops with a *coyote-de-frise* of sagebrush to exclude the rabbits which have multiplied into swarms, so that the farmers pray for coyotes now."

So that the farmers pray for coyotes now. If, somewhere in the American West, there actually were farmers who prayed for coyotes in the late 1880s, their prayers were faint and fell on uncomprehending ears. With the exception of the odd scientist or two, impressed that for some reason this little junior wolf seemed an especially irrepressible creature, in the half century from the 1880s, coyotes had about as many friends in America as did rattlesnakes, tuberculosis, homosexuals, and, yes, Bolsheviks. Farmers, ranchers, writers for any manner of national publications, and eventually employees of almost every state and federal agency involved with wild predators seemed almost to vie with one another in labeling the coyote a vile species of vermin that should not be allowed to breathe up good air. At a distance, the hatred seems hard to square with anything rational. It certainly wasn't based on science and sometimes looked suspiciously like the collateral damage of a puritanical loathing of our own animal natures. But from the 1880s until the 1930s, the received wisdom in America, very rarely questioned, was that the only good American prairie wolf was a dead one. The real question was how to kill as many coyotes as possible in the very shortest period.

For the coyote, who, after all, saw this phase of its story from the wrong end, thousands of years of veneration by Indians had seemed to turn on a dime.

From the time the bison slaughter commenced in the 1820s, it took little more than half a century to clear the Great Plains of that ancient population of animals, which during spans of good weather must have approached 25 to 30 million animals. One effect of that species cleansing was to open up the great grasslands to domesticated animals. Cattle- and sheepmen began taking their herds and flocks into the open-range West in the 1860s, and as the vast lake of bison puddled into a few remnants, ranchers and sheepmen replaced the native animals wholesale with pastoral domestics that they turned into an economy.

Fur traders at western trading posts during this war on the wild fairly soon realized, however, that coyotes had some utility in the global market. Like so many other mammals in the West—beavers, famously, but also bison, elk, deer, otters, minks, muskrats, indeed virtually everything that grew fur on its back—coyotes and wolves wore pelage that interested those involved in the international market in animal skins. Opinions differed about the quality of coyote fur. Meriwether Lewis argued that it was much inferior to fox fur, but George Ruxton thought it "of great thickness and beauty." In a trade where beaver pelts and buffalo robes were worth $3 to $4 or more, a trapped or poisoned, then skinned coyote was worth only a fraction of that. But the skins were much lighter to ship east, and the animals were so common—recall Josiah Gregg's comment that coyotes were "found in immense numbers on the Plains"—that in some parts, coyote pelts began to function as money. Frontier trader James Mead, who in 1864 built a trading post near today's Wichita, Kansas, recorded that to settle a debt of $3,000, the famous trailblazer Jesse Chisholm offered him buffalo robes, wolf skins, beaver pelts, or buckskins. "I chose coyote skins," Mead wrote, "which were legal tender for a dollar, and he counted out three thousand."

Not accustomed to fearing humans, coyotes were not at first in the least wary. Soon enough they learned their error, for by the mid-nineteenth century we possessed a killing agent that did not require stalking or trapping or shooting skills—or even our presence, for that matter—and exploited a predator's willingness to scavenge rather than risk injury in a hunt. Strychnine, made from the seed of an East India fruit, was in commercial production in Pennsylvania as early as 1834. Cheap and entirely unregulated, it became a key tool of biological warfare against the natural world in America for much of the nineteenth and twentieth centuries.

Strychnine was also horrifically deadly in tiny amounts, usually administered in the form of white tablets. Within ten to twenty minutes, a tab of strychnine gulped down as part of a predator's meal wreaked havoc on the central nervous system, launching the victim into waves of wrenching, convulsive cramping—a truly shocking sight when anyone was around to see it, which was rare. Asphyxiation was the cause of death, but strychnine seized living animals with such violence that it left a characteristic signature: dead bodies with rigidly arched spinal columns and straight motionless tails, spotted at distances across the prairies as toppled-over question marks.

By the 1850s, trading posts in Westport, Missouri, where overland travelers struck out across the Great Plains, regularly stocked strychnine for western travelers. This is how predator pelts began to join the international fur trade in that decade. With strychnine pellets in their saddlebags, travelers and traders could lace every dead bison or horse they saw with the poison, hang around a day or two to see what happened, and reap the benefits. Confronting strychnine, western predators were suddenly very vulnerable.

The nadir of the commercial buffalo harvest in the 1860s to 1880s turned the plains into a pathetic slaughterhouse. This created unprecedented boom-time conditions for grizzlies, large scavenging birds, and wild canines, but the time also inaugurated the first of many scorched-earth biochemical wars against such animals. For a few years wolves and coyotes lived large off the blood sport. But as even creatures as numerous as bison began to dwindle away before the market hunt, ultimately hunters had to expand their target species to keep working. So elk, pronghorns, wild sheep, deer, and sometimes even wild horses fell under the gun sights of the grand killing fests of the 1870s and 1880s.

No one, then or now, has ever been able to measure a war on wild things on this vast a scale. There were between 5,000 and 20,000 hunters on the Great Plains in those years, and we have the anecdotes of only a few of them. But western writer and conservationist George Bird Grinnell, founder of *Forest and Stream* magazine, believed that in the case of coyotes, hide and fur hunters killed some "hundreds of thousands" on the Great Plains during those two decades.

Hundreds of thousands is an abstract figure, too big and vague to linger in the mind. But maybe this will. While we'll never get closer to a true figure of all the coyotes killed in those decades of their first encounters with Americans, we can speculate with some certainty that every one of those coyotes wanted to live rather than be shot down, struggle in bewildered fear in a steel trap, or suffer a wretched death from poison. They would all, to use the philosopher Bion's phrasing, have died in earnest.

These western settlers rarely agreed with one another on much, but they did share a hatred of predators. It was an enmity that had begun 10,000 years ago when humans first began herding goats and sheep, ripened in Europe down the centuries, festered in eastern North America and the Southwest when Europeans introduced the pastoral lifestyle there in the 1600s, then reached a crescendo of venom in America after the Civil War.

Cattle ranchers never got too heated up about coyotes so long as wolves lurked about their herds, but sheepmen quickly came to regard them as a "parasite on civilization." Virtually on the heels of organizing stock associations to compensate for losses to predators and rustlers, cattle and sheep raisers began to push for a predator-control tool that New Englanders had used as far back as the 1630s: a cash bounty paid to the man who presented the head or ears of an extinguished predator. The first territorial and state bounties in western America targeted wolves as the primary threat to livestock. But coyotes did not escape notice from ranchers or bounty hunters for long.

Having little initial familiarity with coyotes, new residents often lived in the West for a decade or so before deciding that coyotes deserved a price on their heads. Kansas, for example, bountied wolves in 1864 but did not add coyotes to the list until 1877, when payment for either a wolf or a coyote was set at $1 per "scalp." Colorado Territory created its first bounties in 1869; Montana Territory followed suit in 1883 and Wyoming in 1893. The territory of Arizona and New Mexico created its bounty system in 1893; its list included coyotes as well as wolves, bears, lions, and bobcats.

In the bounty phase of the predator war, Colorado was typical, going for wolves first but adding coyotes in 1876, then raising bounty payments over time, from 75 cents "a scalp" in 1879 to $1.50 in 1881. In 1893 the Colorado legislature began to differentiate between wolf and coyote bounties, paying $2 for the former and $1 for the latter. Urged on by stockmen's associations, whose members tended to dominate western legislatures, by 1914 western states were paying $1 million a year in bounties that overall averaged $1 per animal. It's an easy bet that coyotes formed a plurality of those 1 million bountied animals per year. Not winnable, but easy.

As large apex predators whose presence and domination had always served to suppress coyote numbers, wolves were never as numerous as their smaller cousins, but due to their reputation alone, initially they got the brunt of this lethal attention. A new federal agency, dedicated to the destruction of predators, would soon call bounties into question as an ultimate solution, but bounties undoubtedly produced results, especially with wolves. With packets of strychnine available in every hardware store in America, scattering a few poison tablets across the countryside to help beat back the continent's wild predator horde was almost a patriotic duty for ordinary citizens.

Some governments in the West—in Montana, for example—absolutely prostituted themselves before the ranching industry. Between 1883 and 1928 Montana paid bounties on 111,545 wolves and 886,367 coyotes, a ranching subsidy that grew so large during the territorial stage that it devoured a stunning two-thirds of Montana's annual budget! As a state, Montana bountied 23,575 wolves in 1899, but by 1920 wolf populations had collapsed to such an extent that in that year Montana paid bounties on only 17 gray wolves. Since bountied coyote numbers remained consistent—about 30,000 a year, with no drop at all between 1883 and 1928—in 1905 the state's legislature upped the pressure on coyotes by actually passing a law requiring veterinarians to introduce sarcoptic mange into the wild canine population. This early form of state-sanctioned biological warfare still afflicts twenty-first-century coyotes and wolves in the region.

To many, these measures were not enough, not for wolves and certainly not for coyotes, whose numbers inexplicably remained undiminished despite the extraordinary numbers reported killed. The source of the problem, many westerners came to believe—naturally enough, it had to be so—was the federal government. At the turn of the twentieth century a new federal policy underway in the West represented a sea change, and as it originated with scientists and a handful of eastern intellectuals, westerners were suspicious from the start.

From the time of the first homestead acts, designed by Thomas Jefferson in the 1780s, the public domain that the United States steadily added to the country in the nineteenth century had been administered by the General Land Office, which offered it for sale (or sometimes as free homesteads or grants) to citizens or to infrastructure-building corporations like railroads. For more than a century the public domain's intended destiny, in classic American tradition, was to become private property. Through purchase (Louisiana, Alaska, parts of Arizona and New Mexico), diplomatic agreement (the Northwest), war (the Southwest), and annexation (Texas, Hawai'i), the United States acquired an enormous amount of public domain between 1803 and 1898, and as various federal expeditions explored it, a prescient handful of Americans began to wonder about privatization as a blanket policy aimed so bluntly at such an ecologically diverse range of landscapes.

Two influential Americans in particular worried about this tradition in printed volumes, one a best seller, the other a congressional report. The author of the best seller was American diplomat George Perkins Marsh. A polymath New Englander who read twenty languages and filled diplomatic appointments all over the globe, Marsh in 1864 wrote *Man and Nature*, in effect the first modern history of the environment. Although it took on a huge range of topics relating to humanity's relationship with the natural world, *Man and Nature* became most famous for its discussion of a pattern Marsh had observed in places as disparate as France, Turkey, and China. Rivers, he wrote, had always been crucial to human civilization, and almost everywhere they originated in mountains. But the privatization of mountains, the wellsprings of water that were so critical to human development, had been a disaster almost everywhere that countries had let it happen.

The United States still had time to avoid such a mistake, Marsh believed, by excluding its mountain landscapes from settlement that would invite overlogging and overgrazing and by retaining them instead as public preserves to protect watersheds. Marsh's book went through eight printings and appeared in a new edition in 1871, and its success brought his argument to the attention of the National Association for the Advancement of Science, which in 1873 endorsed this new policy recommendation.

The other author was a one-armed Civil War veteran who became the most famous American explorer of the postwar era and eventually the most powerful bureaucrat in government late in the century. John Wesley Powell had lost an arm at Shiloh, but that could not prevent him from leading the first party to take on the dangerous descent of the unknown Grand Canyon, which he accomplished not once but twice, serializing the account of his adventure in the most popular magazines of the day. Then, in 1878, the year before he became the director of the new United States Geological Survey, Powell laid before Congress his masterpiece for rethinking public domain policies in America. *The Lands of the Arid Region of the United States* didn't exactly endorse Marsh's plan. Powell focused more on the diversity of public domain landscapes and why Congress should tailor settlement plans specifically for valleys, foothills, and mountains. Yet by emphasizing the special difficulties settlers were facing in a West that was far more desertlike than anyplace Americans had ever tried to homestead, Powell added yet another layer of reasoning as to why protecting western water was so crucial.

Neither writer had said the first thing about wolves, coyotes, or any other wild creatures, but in 1891 Republican president Benjamin Harrison's administration passed an appropriations bill for the General Land Office that included a rider placing 13 million acres of western mountains off-limits from settlement. It was the beginning, in the Sierra Nevada and Rocky mountain ranges, of what would evolve into America's National Forest System, a linchpin of Teddy Roosevelt's crucial conservation program a few years later that eventually included a National Park Service and more than fifty National

Wildlife Refuges, all created out of federal public domain lands. By 1907, squat, bespectacled, squeaky-voiced Teddy Roosevelt, rivaled only by Thomas Jefferson as the most nature obsessed of all American presidents, had set aside 151 million acres of western mountain lands as national forests, and by 1910 the new policy began to work on acquiring cutover lands in the East, Midwest, and South as additional public forests. A new National Park Service followed in 1916. By 1932 it was administering twenty-two national parks and thirty-six national monuments, almost all of them fashioned from lands everyone once assumed would be parceled out to private individuals, as American lands had always been before.

Farms, ranches, and towns would be located on the borders of, but not within, these vast expanses of public forests and parks. Despite roads, trails, campgrounds, and tourists, within them big nature prevailed. Instead of replicating the East, settlement in the western third of the country was angling off on a new historical course. For big predators like coyotes and wolves, a historical course that preserved vast expanses of wildlands in America couldn't help but look really promising. Naturally the story turned out to be a lot more complicated.

René Descartes famously thought of animals as so distinct from humans that they were almost automatons, incapable of joy, sorrow, or emotional lives of any kind. According to centuries of Western scientific thinking nearly down to our own time, higher animals' behavioral responses are based purely in instinct, not—like human behavior—centered on morality or advanced emotions like fairness or empathy. Until very recent work by behavioral biologists, we firmly believed that no animal other than ourselves had anything remotely approaching a so-called theory of mind, the ability to recognize that other beings are also engaging in thought and to attempt to discern what other minds outside our own might be thinking.

Most of those who pioneered the systematic, professionalized destruction of creatures like wolves and coyotes in America from the

1880s to the 1930s no doubt regarded wild predators in the Descartes mold. Wild canines were profoundly, fundamentally different from humans. Most, like professional wolfer Ben Corbin, who wrote *The Wolf Hunter's Guide* in 1901, had internalized the Judeo-Christian teaching that humans possessed "souls," which other animals lacked. Corbin was far from alone in imagining this war in terms of a Christian crusade against the depravity of predators. Wasn't morality—at its core, a code for treating others as fairly as we wish to be treated ourselves—an invention of human culture, its precepts disseminated through religious teachings? For centuries the most enlightened position we could muster about our "brute neighbors" was that they deserved some semblance of decent treatment, but certainly not because they possessed emotional lives, experienced pain or loss, or understood anything at all about morality.

Prior to this first phase of America's war on wild things, Charles Darwin's experiences with his own dogs had convinced the great naturalist that canines did indeed possess emotional lives and perhaps even had some fundamental form of morality. The nineteenth-century activists who formed the Society for the Prevention of Cruelty to Animals moved in Darwin's direction, and post-Darwinian writers like Jack London, James Oliver Curwood, and Ernest Thompson Seton attempted early-twentieth-century fictional stories built around Darwin's idea that animals experience joy and anguish and understand the essentials of fairness. But the scientific community quickly and famously squashed such notions. The activists were sentimentalists, scientists of Roosevelt's day averred. As for the writers, not only were they anthropomorphizing animals, trying to turn them into little humans, but they were deliberately falsifying natural history for purposes of sentimental entertainment.

The outcome of what has been known in the history of science ever since as the "Nature Faker Controversy" may have comforted those about to wage a war of extermination against American predators. In our time, though, it no longer works to seek solace in the notion that other higher animals are so different from us that they entirely lack emotional lives. The Stephen Pinkers of the modern world have made us understand that the human senses of fairness, equity, and empathy, the fundaments of the moral code, do not in fact spring from organized religion or advanced culture but have roots in our very evolution as a social species. We are beings with brains that are endlessly taking stock of favors and slights, reciprocity and advantage. Morality did not emerge from religious teachings. Rather, religious teachings encoded a morality that sprang from human social evolution.

That breakthrough in understanding our own animal nature has led, in the present age, to the work of scientists such as Marc Bekoff and Brian Hare, who study primarily the social lives of dogs but also those of wolves and coyotes. Their work leaves little doubt that as social species themselves, canines also understand equality and inequity—morality, in other words—and experience both a rudimentary form of empathy and some basic theory of mind. In practice, theory of mind involves efforts by higher animals to read expressions and body language to discern the outlines of other creatures' thoughts.

In other words, just like us, canines are animals whose evolutionary history as social species has given them an essential sense of what in human terms we would call "right and wrong." Numerous studies have demonstrated that both canines and chimps know when they are being treated equitably among their peers and when they are not, and their behavior registers this knowledge. According to Bekoff, much as humans use prisons and enforced socialization, wild coyotes have a sense of proper coyote behavior and ostracize individuals that fail to observe it. As for theory of mind, domesticated dogs famously demonstrate remarkable canine abilities in common acts that they and their owners perform on a daily basis. When we point at a dish or a toy, with their 15,000 years of coevolution alongside us, dogs read our intent and look in the direction we're signaling. Wolves, coyotes, and chimps demonstrate theory of mind in other ways (often by reading intent in play gestures), but even chimps stare at our fingers when we point. Unlike dogs, they fail to infer the mental signal.

Knowing what we currently understand about the evolutionary origins of human morality and the emotional lives of higher animals, it is difficult to avoid the conclusion that the fate that befell those millions of wild American canines from the 1880s to the 1930s must have caused them some staggering level of emotional trauma and perhaps even a profoundly experienced, rudimentary sense of unfairness, the kind of mental sense that in us becomes a powerfully felt idea: injustice.

In 1897 New Mexico rancher Arthur Tisdale, resident of a territory where the new conservation policies were in the process of creating the Santa Fe and Carson National Forests in the Southern Rocky Mountains, became the first known individual to call on the federal government to assist ranchers facing what was becoming known as "the predator problem." Ranchers like Tisdale reasoned out a syllogism whose logic went like this: Private efforts alone had sufficed to wipe out most of the wolves out on the plains. But because federal policies setting aside new public lands in the mountains *were creating predator refuges* there, having thereby created the problem, it was only right that some government agency help ranchers and farmers clear out the predators that were holing up in preserves precious few westerners had wanted in the first place.

No one either in or outside government investigated this trail of reasoning very closely for evidence. The idea of public lands as animal refuges seemed intuitive. And looking back, there is some good evidence that several species of animals formerly found on the Great Plains—grizzlies, certainly, but also elk and very likely coyotes too—began in the 1880s and 1890s to retreat into the more protective mountains. Always more common on the plains and in the western deserts than anywhere else in the West, coyotes do appear to have begun colonizing westward in the direction of the Sierras and the Pacific in the 1890s. Likely they were pulled there by possibilities around mining activities and pushed there by persecution on the Great Plains.

Since federal administrators of the new federal land preserves, like Forest Service chief Gifford Pinchot, knew they needed the support of locals in the West, and scarcely any of them felt sympathy for predators anyway, the federal agencies responded to Tisdale's arguments. The Forest Service, in charge of the National Forests, thus became the first government agency to kill predators on behalf of ranching interests. Eventually even rangers working for the new Park Service would join the predator war with gusto. Later both the famous Civilian Conservation Corps of the New Deal era and the Grazing Service, the latter created to manage the remainder of the public domain once the Dust Bowl ended homesteading on federal lands in 1934, would make sure that some of the money for conservation and public-lands grazing fees actually went to predator extermination.

But the Biological Survey, an apparently benign federal department designed to continue the natural history work in America that Meriwether Lewis and William Clark had begun, became the federal agency whose very reason for being was predator control. Renowned American scientist C. Hart Merriam planted the seed for today's US Fish and Wildlife Service—and its predator-control stepsister, now known as the Division of Wildlife Services—in 1886 in the form of a federal agency known as the Division of Economic Ornithology and Mammalogy. Merriam renamed it the Division of Biological Survey ten years later.

In the first decade of the twentieth century, Merriam's son-in-law, a scientist named Vernon Bailey who struck many on first meeting as a kindly soul, directed the Biological Survey's ongoing, nationwide cataloguing of the fauna of America. That was the survey's official mission statement up until 1905. Bailey was hardly a St. Francis though—indeed, he would have shot the wolf at St. Francis's side as fast as he could draw on it—and he set the agency on a mission quite different from the one his father-in-law had imagined for it.

Somewhat like today's National Aeronautics and Space Administration, the Biological Survey, due to its original pure science tack, endlessly found itself scraping for funds (and survival) when

Congress passed appropriations bills. By the early century the Biological Survey realized it needed a mission statement that would make it seem a critical, if not downright indispensable, government player. The search for that economic mission kept leading the bureau back to the Forest Service, its National Forest System, and the public image problem it had with western stockmen who protested that the national forests were nothing more than a series of woodsy hangouts for the predators that decimated livestock. To Bailey this seemed an opening, so in 1907 he authored a pamphlet titled *Wolves in Relation to Stock, Game, and the National Forest Reserves*, wherein he argued, among other things, that the national forests *were* serving as predator refuges, and his agency could well be the solution. While he was at it, he expressed the sentiment that although wolves were certainly a menace to sheep and cattle, other, equally worthy targets were holing up in the national forests. He continued, "Wolves kill far less game in the western United States than either coyotes or mountain lions."

Embroiled then in a case that would go to the Supreme Court related to efforts to levy fees for grazing in the national forests, Forest Service chief Gifford Pinchot was in a conciliatory mood toward ranchers. He turned to Bailey to clear the national forests of predators. By their account, Bailey's bureau men proceeded to enjoy an outstanding success in locating predator dens, killing pups, and trapping and poisoning the adults. According to the bureau's next circular, *Destruction of Wolves and Coyotes: Results Obtained During 1907*, his men had dispatched 2,000 wolves and 23,000 coyotes in just a single year. With that outcome, Bailey initiated what would become a long-running campaign against the bounty system, which he argued was an unworkable private and state effort to deal with a problem that really required federal professionals. By the end of 1907, a government bureau that until then had seemed an innocent troop of deer counters was positioning itself as the solution to the "problem of predators."

The Biological Survey needed a clear, pragmatic justification for its existence, and the notion that public lands were harboring and breeding a menace to private enterprise provided its main chance.

The Progressive Era—the years from 1901 through 1916, during the administrations of Republicans Teddy Roosevelt and William Howard Taft and Democrat Woodrow Wilson—represented the real beginning of growth and reach in the US federal government. As with many other projects of this initial age of social engineering, many Americans believed that a large-scale effort like exterminating predators was simply too big a task for individual ranchers. Indeed, most westerners were coming to the conclusion that it was too big even for livestock association or state bounty programs. Coyotes, in particular, for some reason seemed impossible even to thin out. Getting rid of predators called for federal men, experts who understood animals—and who were preparing themselves by training in the techniques of mass killing.

The first congressional "eradication appropriation" finally went to the bureau in 1914: it awarded $125,000 for use "on the National Forests and the public domain in destroying wolves, coyotes, and other animals injurious to agriculture and animal husbandry." Finally armed with the budget and mission it had been seeking for nearly a decade, the bureau hired three hundred hunters around the West to engage in a brand-new, federally mandated war against wild things. Within two years it also asked Congress—and western senators and representatives made sure Congress agreed—to allow the bureau to accept additional private funding from stockmen's associations, as well as money from state legislatures.

At last the Biological Survey had found an argument for its existence that not only brought money rolling in from a variety of sources but seemed to make sense to everyone. That included middle-class Americans of the age, who had internalized Alfred Lord Tennyson's flawed but potent redaction of evolution. "Nature, red in tooth and claw" convicted predators of all manner of crimes and cruelties. Even the Audubon Society endorsed the Biological Survey's antipredator campaign.

There was one other new ally. Public relations experts within the bureau began to mount a campaign to spread the idea to sport hunters that its project of destroying predators, which state after state was now classifying as unprotected "varmints" or "nongame," would have the added benefit of creating bumper populations of "game" animals for hunters to shoot. The idea was that in this brave new world that American wildlife experts were engineering, sport hunters would replace predators in harvesting creatures like deer and elk. It was an absolute stroke of genius. Trappers who had long made money on the pelts of coyotes and wolves didn't welcome the federal competition, but the bureau's argument brought all manner of sportsmen's groups, firearms manufacturers, and state game and fish agencies to the cause.

The Progressive Era was the age of the bureaucratic professional, and professionalism prevailed at the Biological Survey. The quickest, most "efficient" way to mass-kill wolves and coyotes was not shooting individual animals but poisoning entire populations. So with the goal of blanketing river valleys and mountain ranges with poison bait stations that aimed to kill every predator of every species in a region, with its new funding the bureau now proceeded to build a plant in Albuquerque, New Mexico, to produce strychnine tablets in volume. Chillingly and unsentimentally dubbed the Eradication Methods Laboratory, this federal killing facility moved to Denver in 1921, where it would go on to perfect an amazing witch's brew of ever more efficient, ever deadlier predacides. Chemists and researchers in the Eradication Methods Laboratory, with government jobs and benefits, presumably realized the American Dream in the 1920s, buying houses, automobiles, radios, and washing machines, all the latest technologies of the decade. Their products, meanwhile, destroyed America's wild animals, the foundations of an ecology that 20,000 years of evolution had perfected, as if their victims were of no consequence whatsoever.

For the hunters employed by the Biological Survey, the approach in the field was simple. The bureau's professional hunters' first step was "prebaiting," strewing cubes of fat and meat across the countryside to get wolves and coyotes habituated to them. That accomplished, the actual "poison bait stations"—in the age of the automobile, each bait station was commonly one of America's surplus horses, which could be led to the selected spot and shot and whose carcass was then laced with strychnine tablets and surrounded by poisoned fat and meat cubes—went in next.

Stanley Young, one of the bureau's initial hunters who rose to subsequent prominence in the agency, became something of a coyote specialist in this new game. Young had grown up in Oregon idealizing Lewis and Clark. Now, in places like the Sacramento Mountains of New Mexico and along the rims of the Grand Canyon in Arizona, he discovered that with strychnine it was possible to kill 350 of Lewis and Clark's "prairie wolves" in less than ten days. Approaching his bait stations, he later wrote, he found that he could tally a quick total of his victims even from a distance because of the way they died. Every single dead coyote was frozen into that signature strychnine convulsion—a wrenched, alien shape easily visible against the landscape, its tail sticking straight out and frizzed as if the animal had been struck by lightning.

Young's visual imagery of the US government's coyote extermination campaign was soon writ large across the West. With wolf populations rapidly collapsing in the face of the bureau's war on the wild, señor coyote's turn was coming fast. But there remained one very large and visible public-lands arena in which wolves were still the main target and coyotes still mostly collateral damage in the wolf war. That was the national parks.

Yellowstone, set aside as the world's first national park in 1872, and Glacier, created along the Continental Divide in Montana in 1912, became symbolic national scenes of America's wolf and coyote jihad in the 1920s. The United States created national parks in order to allow the public to experience wild nature in its pristine state, so you would assume that when Secretary of the Interior Henry Teller banned all hunting in Yellowstone, even predators would have found refuge. But naturally, park managers saw things differently. Despite having been on the "pristine" Yellowstone Plateau for 1 million years before the park ever existed, wolves, lions, bears, and coyotes somehow were unwelcome enough to produce extermination campaigns even in America's grand nature preserves. Sport hunters, hoping the parks would breed game animals, and ranchers, hoping they wouldn't become asylums for predators, pushed for this, but they didn't have to push hard.

Yellowstone aimed its first predator-extermination campaign directly at the park's "numerous and bold" coyote population. At first the army rangers who patrolled the park randomly shot every coyote they saw. But as early as 1898, park personnel actually began to poison coyotes, mountain lions, and wolves inside park boundaries. When Congress signed the Biological Survey's predator death sentence appropriation in 1914, Yellowstone went so far as to invite Vernon Bailey to show park personnel the proper mass-extermination techniques. Following Bailey's approach, which focused heavily on dens and pups, between 1914 and 1916 Yellowstone rangers destroyed eighty-three coyotes and twelve wolves inside park boundaries. Stephen Mather, the charismatic New Englander who became the National Park Service's first director in 1916, is a conservationist hero to many, but Mather thought that if a Yellowstone ranger "didn't kill off his 200 to 300 coyotes a year," the park's coyotes would spread across the West and wreak havoc. Yellowstone's tally until the death of the last wolves in the park, in 1926, was 136 gray wolves, 80 of them puppies. Coyote deaths, of course, did not end in 1926. With bureau assistance, the number of coyotes that died in Yellowstone from 1918 until 1935 reached nearly 3,000: 2,968, to be precise.

Glacier National Park, 250 miles north of Yellowstone, played a tail-to-a-kite role to the older park, primarily because its clusters of rounded mountains and glacial valleys did not provide as rich a habitat for herbivores or predators as Yellowstone. Yet James Galen, Glacier's superintendent in 1913, thrilled by the state of Montana's experiment with biological warfare against predators, wrote the state veterinarian, "I am desirous of inoculating, with mange, some coyotes to turn loose here in the park, with the idea that I may eventually kill off all the coyotes in the park in this manner." Montana dutifully supplied a pair of mange-ridden coyotes to Glacier in late 1913, and

Galen turned them loose with best wishes for success. But Department of Agriculture bureaucrats in Washington eventually scotched his bigger plan to spread mange to wild canids. Poison, they thought, would be far more effective, especially if the baits were placed on the border between Glacier and the Blackfeet Reservation, a "breeding ground for coyotes." As for wolves, Glacier killed only fourteen between 1910 and 1920, although the park's proximity to a healthy Canadian wolf population made it a sporadic wolf colonization destination throughout the twentieth century.

All the government's hunters, whether on private ranches, in the national forests, or in the parks, initially concentrated their greatest efforts on wolves because, frankly, the war was all about the interests of the livestock industry, and ranchers particularly hated wolves. But within a decade after the Biological Survey's mandate, its hunters had so thoroughly reduced the gray wolf population that after about 1926 the bureau's hunters rarely killed more than a single wolf a year in any state. Nonetheless, in 1923 in the single state of Colorado, the bureau had set out 31,255 poison bait stations. This was the start of a new phase. With almost no wolves left alive in America, coyotes had now become public enemy number one.

In fairness, bureau explanations for an increasing focus on coyotes were not entirely matters of expediency to preserve funding after the poisoning campaign against wolves turned out to be a little too successful. With the country's keystone predator now gone, an ecological chain reaction set in across much of America. Their 20,000-year competitor canines now almost erased, coyotes began to exercise their ancient fission-fusion capabilities. Some coyotes began to form packs and hunt larger prey, including sheep and occasionally (although very rarely) calves. Either as pack members or in singles and pairs, coyotes proved far more elusive to federal hunters than wolves, whose social bonds were so strong that pack members tended to fall one after another into traps baited with the scent of their pack mates and puppies.

Coyotes were also on the move. With wolves disappearing, coyotes of the 1920s and 1930s found themselves in a world where

The federal coyote killing program exists as a subsidy for a fading US sheep industry.
Courtesy Dan Flores.

humans seemed their only threat, but whose forest clearing, cities, and built environments also offered coyotes brand-new opportunities. Harassed and endlessly pursued, but now by people rather than wolves, coyotes in the 1920s ratcheted themselves into survival high gear. Not only did they employ several then unsuspected evolutionary stratagems for maintaining their populations in the West, but they began another historic expansion of their range—first westward toward the Pacific and northward in the direction of the Yukon and Alaska, then eastward across the Mississippi River into the East and South, where they would gradually begin to fill a niche left almost entirely vacant by eradication campaigns against wolves.

With the wolf cleansed and the bureau's Denver plant cranking out the strychnine, by the mid-1920s bureau hunters reached the rather phenomenal milestone of having set out 3.567 million poison bait stations across the West. This scorched-earth policy against coyotes yielded some 35,000 dead coyote bodies a year, although the bureau publicly estimated that its hunters never found another 100,000 poisoned annually. But soon enough it began to dawn on the bureau's overconfident operatives that even after they had saturated the landscape with poison, coyote numbers somehow weren't diminishing. Like Old Man Coyote in the native traditions, the real coyote simply refused to die. First surprised, then increasingly angry, bureau personnel began to confess privately to one another that for a reason no one could figure out, a bureau with wolf notches on its gun was struggling to win the war of civilization against the slinking, lowlife junior wolf.

The bureau's success against wolves combined with hubris to blind its functionaries, at first, to an observation that had actually been around for a while. The truth was, people who had experience with coyotes had been puzzling since the late 1800s over their extraordinary resilience. Unlike so many of the West's animals—bison, bighorn sheep, antelope, elk, grizzly bears, and ultimately wolves—all of which had turned up their toes in capitulation to American western expansion, coyotes seemed an anomaly. And as early as 1900, one of America's leading national magazines had published a short story by a famous nature writer that directly addressed the coyote's singular situation in America's war on wild things.

Ernest Thompson Seton was a Canadian who ended his long and productive career as a writer (and founder of the Boy Scouts) living in New Mexico. Seton has taken his licks across the years as a fellow traveler of the "Nature Faker" writers of the early twentieth century, the group famous for anthropomorphizing the animal characters in their books. In some of the nature writing of that time, the animals reasoned and had morals, societies, and advanced cultures with laws. One critic of Seton's book *Animals I Have Known* even wrote, sarcastically, that its proper title ought to have been *Animals I Alone Have Known*.

Nonetheless, Seton's most famous coyote story, "Tito: The Story of the Coyote That Learned How," which was the lead piece in the August 1900 issue of *Scribner's*, had taken up coyote resilience back at the turn of the century and tried to explain it allegorically. Like the Indian stories featuring Coyote, Seton's "Tito" took on an observable truth and offered an explanation. Species after species was disappearing in twentieth-century America. Yet, despite a "fierce war" that "had for a long time been waged against the coyote kind," coyotes somehow had not done the proper and expected thing. For some reason, they had refused to disappear.

To offer an explanation, Seton invented "Tito," a little bobtailed female coyote captured as a pup and chained in a ranch yard as a curiosity. But she was observant and coyote-smart, and this close association with humans taught her not just about the range of dangers from humans but how to avoid lassos, metal traps, gunfire, and poison bait. From experience as an "insider," she learned to hoodwink hounds and finally grasped the ultimate trajectory of man's designs against "the coyote kind." Tito had made a lousy dog, so was never a pet, and ultimately she escaped, found a mate, had a litter of her own, and then proceeded to teach her pups and "their children's children" all the wisdom she had learned.

In the story's formulation, through transmission down the generations from a Hero Coyote, all coyotes after her would be "wise in the later wisdom that the ranchers' war has forced upon them." Seton's human analog to Tito? Moses, of course, who by growing up among the Egyptians learned their ways, which enabled him to save the Israelites from Egyptian slavery and persecution.

So just how were coyotes able to survive a nineteenth-century fur trade followed by a twentieth-century war of extermination without going under? As Seton's story implied, intelligence and the sharing of learned behavior certainly enabled such coyote resilience. But in truth, the coyote's evolutionary biology played as large a role. The fission-fusion flexibility that went so far back in their evolution made coyotes—very much like us—opportunists able to thrive in wholly new circumstances. In our own case, fission-fusion abilities early in our evolution allowed us to survive bottleneck die-offs, probably from disease epidemics that threatened to exterminate us. In the case of coyotes, fission-fusion helped position the species to survive twentieth-century America's war on predators when wolves could not. A primary reason coyotes are in Los Angeles, Chicago, and New York City this morning has to do with an evolutionary adaptation they share with us.

Coyote evolution also produced a litany of other traits that help explain the species' resilience, some of which should appear familiar to us as well. Like us, coyote young have lengthy childhoods during which they learn from their parents cultural skills and critical information about the world. Juvenile coyotes, like human children, have to be taught through their adolescence the cultural wisdom of their species. So Seton was right about Tito, to a certain extent. Our intuitive grasp of this aspect of the social life of canids is one reason wolves were our first and oldest domesticate. We can also relate to the inclination of wild canids like coyotes to pair up as long-term mates and for both parents to assist in rearing young. These are not especially common mating strategies in nature, but they seem normal to us.

Decades after Seton's story appeared, though, biologists discovered another highly important coyote adaptation that helps explain their resilience, and it amazed even the scientists. The average coyote litter size is 5.7 pups, but that number can range from as low as 2 to as high as 19. The reason for such variability is that coyotes possess an autogenic trait that allows them to assess the ecological possibilities around them. If not persecuted, they saturate a landscape to carrying capacity, then usually have small litters that produce only a couple of surviving pups. But if they sense a suppressed coyote population relative to available resources, they give birth to very large litters. The coyote's yipping howl, known around the world as the iconic music of wild North America, has several functions, one very important one of which is to assess the size of the surrounding coyote population. In the face of persecution that thins their numbers, they respond with whopping litters with high pup survivability.

Other evolved traits also enable coyotes under stress to maintain or even grow their populations. Coyote packs commonly consist of a breeding alpha pair along with one- and two-year-old pups, often females, from previous litters. If something kills the alpha female, lower-ranking beta females, which can breed as early as ten months, will come into estrus during the normal late-winter mating season and have litters of pups as, in effect, adolescent mothers.

Coyotes are remarkable in other ways too. Fission-fusion flexibility makes them omnivorous, and they may scavenge, but nature designed them to succeed as predators. We often express horror at animals that pursue and kill other animals, but such a response demonstrates a misunderstanding of our own evolutionary history. We have been a wildly successful species in part because of our predatory skills. We ought to look with clear eyes and admiration at the coyote's skill set. Darwinian evolution gave its species something like 1,000 times the number of scent receptors we have. Its ability to hear extends into extremely high frequencies in the range of 80 kHz, about 25 percent higher than dogs. As is the case with many other species, a coyote's eyes lack cones to separate wavelengths into color. Bureau hunters trying to eradicate them got scant advantage from those yellow-orange coyote eyes perceiving the enemy in black and white though. Coyote vision is at least as good as ours, and their peripheral vision is much better.

Along with cultural transmission that instructs pups about how to be predators, stalk a deer mouse or vole and pounce on it with stiffened front legs, or suffocate larger prey with a bulldogging neck grab and a bite hold that collapse the windpipe, coyotes bring to bear on the world great observational learning intelligence. Occurring for 20,000 years alongside wolf evolution, coyote evolution selected for animals that were naturally wary, even nervous, traits usually associated with a species that is itself sometimes prey (again, behavior we ought to recognize in ourselves). In fact, coyotes are not "cowardly" but instead are circumspect to the point of extreme wariness. Lacking a fearsome predator until we arrived, gray wolves possessed a comparative boldness that rendered them relatively easy to extirpate in the twentieth century, whereas coyotes were not. Yellowstone wolf biologist Doug Smith once put it to me this way: "When you're top dog in evolutionary history, you get bold and cocky . . . but when, like the coyote, you've been persecuted your entire existence, you learn how to be clever."

A coyote's nervousness makes it highly suspicious of new developments, new objects, and new smells in its habitat, and it learns extremely quickly from experience. Here is one way I know that. Coyote packs establish fixed ranges that in rural areas may be as large as ten to twelve miles around a den site, territories they defend against other coyotes, scent-mark with urine, and navigate via routinely traveled trails. In the 1980s I was building a house in a canyon called Yellow House out in West Texas, and over coffee in the mornings I began to notice an alpha female coyote and two yearling pups traversing the same trail, day after day, along the edge of a mesa about a hundred yards from my construction. Every morning, at a particular clump of yuccas, the adult female would pause, hump her back, extend a rear leg toward her nose, and scent-mark.

One day, curious about what would happen and maybe feeling a little perverse, I walked over to the same yuccas and relieved myself in her spot. Early the next morning, as usual, there she was, trotting with lolling tongue along her footpath, pups fifteen feet behind. This time, though, at her yucca scent-mark she stopped cold, extended a pointed nose, sniffed, and sniffed again. Then she walked stiff-legged around the clump, a series of steps the pups mimicked exactly. I worked on my house for several more months, and once or twice the pups came and sat on their haunches in the yard fifty feet away and watched in fascinated wonderment. But I never saw the alpha female on that mesa trail again. I'd introduced something new into her routine, and she was too wary to tolerate it even once.

So the coyote's innate wariness, the kind of nervous brainpower that emerges in an animal that is both predator and prey, made it a far more formidable target than Biological Survey bureaucrats ever realized. Coyotes could call on the perfect suite of traits to survive a war against them when wolves, bears, and almost no other North

American animal could turn that trick. Many of the coyote's resilience traits were actually adaptations to thousands of years of living alongside wolves. But ultimately human harassment triggered the same set of hardwired responses persecution by wolves had. Pressuring coyote populations by killing large numbers of them kept the species in a constant state of "colonization," employing fission-fusion strategies to enlarge their prey base, attempting to grow their population with larger litters, and raising a higher percentage of pups to adulthood because there was more food in a landscape where coyote numbers were suppressed.

Tito made a lovely analogy to Moses, but considered all together, these factors explain why coyotes evaded the slaughterhouse we had prepared for them. Try as we might to destroy them—and in this we were stubbornly, obsessively, blindly happy to indulge—they responded with maddening nonchalance.

By the late 1920s, when the bureau finally designated the Division of Predatory Animal and Rodent Control to specialize in its professional killing program, the total eradication of coyotes had come to seem the only reasonable policy in America. Ernest Thompson Seton's fellow nature writer John Burroughs had argued as early as 1906 that predators "certainly needed killing." The "fewer of these there are, the better for the useful and beautiful game." William T. Hornaday, a conservation hero who saved the last of the bison and led the charge to replace commercial hunting with sport hunting, insisted that where predators like coyotes were concerned, "firearms, dogs, traps, and strychnine [are] thoroughly legitimate weapons of destruction. For such animals, no half-way measures suffice." Not even John Muir, who found coyotes "beautiful" and "graceful," mounted a campaign against the destruction of predators, although he worried in print that slaughtering coyotes would induce a "penalty for interfering with the balance of Nature."

But in the 1920s, almost out of the blue, coyotes began to acquire champions. At their annual meetings beginning in 1924, the American Society of Mammalogists began to debate whether predators might actually serve some essential functions in nature and whether American poisoning policies were drastically wrongheaded. A cadre of famous scientific luminaries soon spoke out against the bureau, to the shock of many of its employees, biologists themselves and their critics' colleagues.

In reaction to this new reflection from the scientific community, the bureau would double down on its denials of a role for predators and propose a shocking final solution. In the late 1920s it had a predator war budget of nearly $2 million from its various sources, yet repeatedly conveyed to Congress that it was underfunded in so herculean a task. Those repeated pleas led Texas representative James Buchanan to wonder aloud if a significant increase in the bureau's budget might not enable it, finally, to "wipe out" coyotes entirely as it had done with wolves.

Ignoring signs of growing restlessness within the scientific community, in 1928 the bureau's Major E. A. Goldman, its point man on all things canid, offered up the agency's predator endgame. If Congress would fund the bureau at $10 million for a decade, it would wipe out coyotes completely, once and for all. The bill the bureau drew up and sought out Congressman Scott Leavitt of Montana to sponsor was not to be called the Coyote Extermination Act, however. Even in the 1920s that sounded tin eared. Internally, the bureau called it the Ten-Year Bill. If enacted into American law, it would be called the Animal Damage Control Act, a fantasy law for the agricultural industry and for the bureau too. The figures involved were certainly fantasy: the sum of $1 million a year for ten years to effect the outright "eradication" of coyotes appears to have been drawn out of thin air, designed more for effect in Congress than anything else. But even with coyote numbers still high and science beginning to turn against it, with the Ten-Year Bill the bureau did actually seem to believe that a civilized kingdom in America was at hand.

It had been only 125 years since Lewis and Clark first encountered "prairie wolves" and wondered what to think of them. Now bureaucratic coyote killers were salivating at the chance to wipe 5 million years of North American canine genetics from the face of the Earth. What could have been more American?

Standing on my patio in the ancient coyote range outside Santa Fe, with coyote howls enlivening almost every night in the High Desert of New Mexico, I wonder now at the mind-set of it all, the mode of thinking that makes us such stoic killers, able to extinguish with such ease the very qualities that lend the world beauty, grace, romance. And, to my personal regret, I wonder not just how it played out a century ago or in other minds. I wonder about it in all of us, certainly in myself, who as a teenager, when coyotes were colonizing Louisiana, tried so hard to possess wild nature and bring it to hand.

This is an uncomfortable memory for me, but here it is. It is an early daybreak, with the sun a flattened red ball through the mists of the Red River Valley. I am seventeen years old. A coyote pauses in yellow prairie grass, her muzzle wondrously sharp and refined, her ears working. Dew droplets cascade into silvery pearls in the air above her as her tail switches the grass. Her intense eyes bore straight into mine: she is posing an ancient question, one I will not be capable of answering correctly until another decade of living has passed. So a rifle blast shatters the humid morning air, and she yelps, spins, disappears.

The next moment is one of the most vivid mind's-eye pictures of my life, as perfect in my memory as a circle. The sun suddenly breaks through the mist, and all that only an instant before had seemed wild, romantic, beautiful, dissolves in stop-frame motion as I look on. The "prairie" becomes a scraggly pasture littered with cow dung and discarded plastic soft drink bottles and broken farm machinery. The "wilderness" is now encircled with a half-collapsed barbed wire fence decorated with rusted, bullet-riddled no-trespassing signs.

And somewhere beyond the cottonwoods along the river, the gears of a propane delivery truck are grinding like chalk on slate.

In an instant I had personally recapitulated the last two hundred years of coyote history. I had destroyed what I loved, drained beauty and perfection from the world with a syringe as I looked on. Detached, stoic. A killer.

CHAPTER 4

The Archpredator of Our Time

The men who graduated from their rabid pursuit of wolves to become trappers and poisoners of coyotes in the 1930s and 1940s were also detached, stoic killers, but they were grown men drawing a salary for their efforts, not teenage boys romantic about the American West. Interestingly, though, they did romanticize the animals they executed. Federal wolf hunters and the stockmen in whose interests they worked were known in those years for anthropomorphizing their targets, giving them names and personalities. The Bureau of Biological Survey's celebration of the craftiness and worthiness of their canine opponents seemed almost to echo the Indian Wars.

As obvious as the psychology of that was, it nonetheless bequeathed to us stories of a last few individual wolves—Rags, Whitey, Lefty, the Greenhorn wolf, the Custer wolf, and female wolves the hunters named Bigfoot and Unaweap. Then there was the famous female Three-Toes, so desperate to find a mate in a now wolfless Great Plains that she eventually seduced and mated with a collie. Having betrayed humans by going feral, Three-Toes's domesticated paramour soon enough followed her (along with their wolf-dog hybrid pups) to his death at the hands of bureau hunters.

By 1923 the possibilities for pursuing heroic, last-stand wolves in the Lower 48 were virtually over. It was at this point—in moves just as psychologically transparent as before—that the Biological Survey first feigned shock at, then resignation to, just how frightful and destructive a predator *the coyote* actually was. The truth was that, with wolves gone, the bureau badly needed coyotes to serve as uberpredators for the purpose of keeping the agency alive. But it was also true that no one at the bureau ever made the connection that wholesale eradication of gray wolves was removing the coyote's sworn enemy of the past 20,000 years. In modern Yellowstone National Park in our own time, we have had ringside seats to this process working in reverse, as gray wolves have rejoined coyotes in places like Yellowstone's Lamar Valley. But back in the 1920s, as wolf after wolf disappeared, the coyote nation found itself in a new world where it had become target number one, slated for official extinction at the hands of the American state.

So the lead-in to the shocking Animal Damage Control Act of 1931 was a wolfless 1920s America. American policymakers have always needed enemies, and with wolves gone, the coyote stepped unsuspectingly into the glare of a very intense predator-hatred spotlight. Suddenly (or so the bureau asserted), cattlemen who had paid little attention to coyotes before realized that "heavy losses of calves, heretofore attributed to wolves have evidently been due to coyotes"— which were now, shockingly, more visible in America than anyone could remember. Whether coyotes really were filling the niche of wolves, or whether many of those stock losses (as scientists who studied the matter in the 1930s believed) were exaggerated or actually caused by feral dogs was not really a matter of science, since a bureau that had once been a vehicle for pure science now only devoted 3 percent of its budget to scientific study.

In this vacuum of reliable information, the coyote assumed the mantle, in a phrase the bureau would use soon and often, of "the archpredator of our time." Even some of the romance got transferred. To increase coyote worthiness as a bureau opponent, distinctive coyotes began to acquire names, such as the "Rick Creek coyote" in Colorado or "Old Crip," a female in Texas that, finally trapped in 1944, supposedly drowned herself rather than be taken. Among themselves, in their own packs and even among other animals inhabiting their world, coyotes were always individuals, as distinctive from one another as we are. Now the humans who pursued them began to distinguish individual coyotes, but in a move that cannot have made any sense to wild coyotes at all, their human enemies began to bestow on them both personalities and motives that oddly resembled those of fascist figures with designs on the American way.

In our twenty-first-century world, the terms "genocide" and "ethnic cleansing" sit uneasily in the mind, associated with some of our darkest and most disturbing thoughts about human nature. They conjure Darfur, Serbia, Cambodia and Pol Pot, and, most vividly of all for many of us, the horrors in Europe before and during World War II. "Species cleansing," on the other hand, is not a term that falls readily to hand, although we have engaged in it without much remorse for at least 10,000 years and probably more. Be it North American mammoths, driven to annihilation ten millennia ago by bands of a near-professional hunting culture known as Clovis, to flightless birds, clubbed and battered to extinction across the islands of the Pacific when Polynesians and later Europeans arrived there, to passenger pigeons and ivory-billed woodpeckers and Carolina parakeets in twentieth-century America, humans are ancient veterans of the art of species cleansing, the act of pushing fellow animals into black hole oblivion.

When Henry David Thoreau, lamenting the phenomenon in 1856, wrote that he did not like to think that some "demigod" had come before him to pluck from the heavens the best of the stars, that he "wished to know an entire heaven and an entire earth," he was mourning a deep time human activity that likely extends back as far as the epoch of Lascaux and Chauvet caves. Even so, few if any organized states have ever been so coldly calculating about species cleansing as to set into law a statute largely conceived as a strategy to exterminate a singular mammal native to its continent. If residents of Queens or the Upper West Side want to know why coyotes are sleeping in their flowerbeds or peering down from the rooftops of bars, the interrogation can begin here, with twentieth-century America's Dr. Strangelove designs to eradicate the animals wholesale.

It is also the primary reason why, two decades after American ecologists first organized to begin mapping out the study of Darwinian relationships, the ten-year plan for coyote eradication in the form of the Animal Damage Control Act of 1931 became a line in the sand for many of the scientists of the period. What ensued from 1931 to 1950 between science and federal policy amounted almost to a predator-prey dialectic. On one side was a scientific community becoming convinced that a federal species-cleansing program for coyotes (and predators in general) was a stunning, myopic mistake, without scientific basis, carrying with it profound collateral damage to nature. On the other was a government bureau, with almost frenzied support from livestock associations, the Farm Bureau, and legislators from the rural West, determined to seize its main chance with a witches' brew of poisons stimulated by World War II. This forgotten war in American history was epic. And unlike other wars of the era, it was one we lost.

From the perspective of the coyote going about its usual rounds, finding mates, establishing territories, and forming packs to enable alphas to raise up new generations of pups, its status as archpredator presented both danger and opportunity. Shot at on sight, run down with cars, trucks, and dogs, and endlessly tempted with easy treats that disguised mortal danger in the form of traps or poison, coyotes in the early twentieth century found themselves pushed hard to explore new chances in a modernizing world.

Acquiring human champions meant nothing to coyotes, but they were materially affected by the fact that since Europeans had arrived, scarcely any humans spoke well of them (save traditional Indians who still credited Coyote with the creation of North America). Along with Ernest Thompson Seton, a usual suspect in such matters was John Muir, founder of the Sierra Club and of a kind of philosophy of American nature worship that endures in activist environmentalism today. But like everyone else at the time, Muir knew little of predator ecology and nothing of the indigenous Coyote lore of his adopted state of California. His Colorado compatriot as a nature writer was Enos Mills, like Muir not a sportsman but a strong advocate of a new idea: the virtue of being in nature without a gun. Gun in hand, Mills had found coyote howls "menacing." But weaponless in the wild, not thinking of coyotes as targets, he realized that howling coyotes were actually playful, full of merriment. Eventually he concluded that as the scourge of mice and gophers, a coyote "does man more good than harm." Living with them high up in his mountain valley below Long's Peak, Mills came to believe that "wise coyote" knew more than we newly minted Americans ever suspected.

It shouldn't surprise us now that the first group of Americans after Indians finally to "get" coyotes were scientists. Elliott Coues, a frontier scientist who spent considerable time in coyote country, expressed an initial lukewarm admiration in 1873, the year after Mark Twain's description in *Roughing It*. America's unique wild canid, Coues wrote, "theoretically compels a certain degree of admiration, viewing his irrepressible positivity of character and his versatile nature. If his genius has nothing noble or lofty about it, it is undeniable that few animals possess so many and so various attributes, or act them out with such dogged perseverance."

Understanding of the role predators like coyotes played had its beginnings when scientific naturalists formed the Ecological Society of America, which met for the first time in 1914. America's founding ecologists, Frederick Clements, Charles C. Adams, and Victor Shelford, agreed at that gathering on several basic strategies for their field, among them the study of adaptation that had been so critical to Charles Darwin's insights, an investigation of the flow of energy through nature, an analysis of "climax conditions" (which fascinated Nebraskan Clements), and development of better insights into how humans disturbed the natural world. Shelford, who had published the landmark *Animal Communities in Temperate America* just the year before, pushed his fellows to recognize and work on biotic communities too.

But the most old-fashioned research topic of all—an idea Western culture had known since the time of Aristotle as "the balance of nature," the presence of a dynamic equilibrium in the natural world—began to push ecological science in the direction of understanding the role of predators. The Biological Survey's policies assumed the European folk position: predators were entirely disposable, and the banishment of wolves and cougars and coyotes from America would create a civilized paradise for deer and elk and ranchers and sheepmen. This thinking ultimately became the rapier point of scientific inquiry.

The man who would become the most famous ecologist of this era, Aldo Leopold, would admit that all the way up until the early 1920s, he had thought in Elysian Fields terms himself. But years later, reviewing the book *The Wolves of North America* by bureau stars Stanley Young and Edward Goldman, Leopold wrote that he had come to realize that a predator-free "paradise" contained a fatal non sequitur. How had it happened that the wolf and coyote population had failed "to wipe out its own mammalian food supply" millennia before Europeans had ever come to North America? Between 1914 and 1945, Leopold's colleagues had studied their way to an understanding of the balances that had kept American ecologies healthy for century after century without human intervention. But somehow, Leopold wrote, the bureau and men like Young and Goldman had obstinately refused to hear this rather self-evident message.

One of the most prominent names in conservation in the decades on either side of the turn of the twentieth century was Grinnell, largely because of the radiant halo cast by George Bird Grinnell, a first-rank conservationist in an age that produced Teddy Roosevelt and Gifford Pinchot and the public-lands system that was their legacy. Grinnell had an illustrious career, founding the Audubon Society and becoming (along with TR) a charter member of the Boone and Crockett Club and later of the American Bison Society, which helped bring bison back from the brink of extinction. He was best known in his day for editing the prototype outdoor magazine, *Forest and Stream*, and for almost single-handedly promoting the Continental Divide region of northern Montana as a new national park, 1910's Glacier National Park.

The glow of Grinnell's fame inspired other members of his large family, and among them was a younger cousin, Joseph Grinnell, who in 1916 would advance the "big idea" that provided ecologists their wedge issue as emerging critics of the bureau's predator policies. Raised on western Indian reservations by his physician father, Joseph Grinnell grew up to become one of the West Coast's most celebrated naturalists. In 1908, after a stint at Caltech, he got the University of California's appointment as the first director of its Museum of Vertebrate Zoology in Berkeley, where he pioneered field techniques for the most careful and thorough collection of mammals and birds ever assembled for a state. As an original thinker, though, Grinnell made his major contribution to ecology with his proposal, in 1924, of the ecological niche, a fundamental insight into nature.

The ecological niche breakthrough was critical for understanding wild coyotes and appreciating predators generally. In nature a "niche" is analogous to an occupation in human culture. As with doctors and dentists in rural regions of the human world, niches in nature sometimes go unfilled or can become vacant. In the case of wolves, coyotes, and mountain lions, an understanding of niches caused ecologists to worry about the result of vacancies. As Grinnell and his students were able to demonstrate, niches for predators had existed across time as part of the balance of nature that ancient societies had marveled at.

In scientese, the coyote trotting across America and going about its ancient business occupied the niche of a midsize predator. It helped keep the populations of everything from mice to some ungulates in balance. The niche of the coyote-sized predator is almost universal across the world. It is filled by wild dogs in Africa, jackals in southern Europe and the Middle East, and marsupial carnivores on the great island continents of the Southern Hemisphere. The Tasmanian tiger was the equivalent of the gray wolf on that island, while the Tasmanian devil—surviving now only on Tasmania—was Australia's coyote-sized carnivorous marsupial and occupied the same niche as a coyote. Humans introduced wild dingos to Australia 4,000 to 5,000 years ago, and with the extirpation of the marsupial carnivores, these true canids then occupied that ancient niche.

The famous Kaibab Plateau deer episode in the 1920s furnished ecologists with a dramatic and helpful story for illustrating how predators create a balance in nature. By then bureau and bounty hunters had managed to erase wolves and mountain lions from the north rim of the Grand Canyon, and for good measure they had poisoned nearly 8,000 coyotes there. The evident consequence in the mid-1920s—exactly when scientists were first challenging the bureau's predator policies—was a population explosion of mule deer on the Kaibab Plateau from roughly 4,000 to 100,000 animals that destroyed their browse and then suffered a catastrophic 60 percent die-off. In a much publicized (and ridiculed) move, Arizona novelist Zane Grey organized a nature-loving group, which included Hollywood actors, that attempted to drive the surviving animals to a new range. The main result was that Kaibab became a national story.

While modern ecologists have questioned the simple conclusions contemporary scientists drew about Kaibab, at the time few ecologists looked for more nuanced explanations. That was especially true when after Kaibab, in 1927, a rodent population explosion in Kern County, California, left highways grossly slick, and ultimately undriveable, after traffic flattened unbelievable swarms of mice. That event also came on the heels of mass coyote poisonings. The lessons of Kaibab and Kern seemed so clear at the time that they served as evidence for the so-called Lotka-Volterra equations of that decade, algorithmic ecological models of how prey and predator populations follow an oscillating rise and fall of first the hunted, then their hunters.

His niche insight gave Grinnell enormous gravitas in ecology, but it was only the opening to the proposal that led his fellows to begin their break with the bureau. Grinnell was also an ardent proponent of the new federally held public lands of America. Although by World

War I America had established sixteen national parks since 1872, there was as yet no managing federal agency in charge of setting policy for America's parks. But when the San Francisco earthquake and fire of 1906 resulted in the damming of Hetch Hetchy Canyon, a large and scenic portion of Yosemite National Park, as a water source for the city, the anguished outcry from preservationists finally produced the creation of a National Park Service. The year was 1916.

So there was now a National Park Service and an organic act that empowered it to preserve nature in the parks for future generations. In terms of policies to preserve nature, though, exactly how would the new Park Service manage these crown jewel American landscapes? Joseph Grinnell and one of his zoologists, Tracy Storer, laid out a suggestion in "Animal Life as an Asset of the National Parks," an article they cowrote for the journal *Science* that same year.

What Grinnell and Storer suggested was radical given what was already happening to predators in Yellowstone and Glacier. "As a rule," they wrote, in the parks "predaceous animals should be left unmolested and allowed to retain their primitive relation to the rest of the fauna." Presumably national park superintendents would be reading *Science*, and the two had a message. They were "naturalists," they wrote, and as such were convinced there was a longstanding balance of nature relationship between predators and the local game animals. No worries, in other words, about sacrificing game to predators. Besides (they went on), "many of the predatory animals" were themselves "exceedingly interesting" to observe.

The idea of offering wolves, mountain lions, and coyotes permanent refuge inside America's national parks, where scientists could do "research in natural history" on them, was shocking at the time. In 1916 the bureau was still a decade away from reducing wolf numbers to single remaining animals like Three-Toes, and the Animal Damage Control Act was still fifteen years in the future. But Grinnell and Storer could see the direction bureau predator policies were taking. At least, they said, let's do what we managed to do for iconic native animals like bison and make the parks refuges for both game animals and predators, places where the grand cycles of primitive America could remain forever intact.

This "big idea" would be debated at meetings of the brand-new American Society of Mammalogists, founded in 1919, and in the pages of its publication, the *Journal of Mammalogy*. After a decade of success and widespread praise for their efforts, bureau leaders were gobsmacked to find that the scientific community was having doubts about what seemed to them nothing less than a mission of civilization. Some of the early stars of the bureau from the days when it had been a purely scientific agency, like founder C. Hart Merriam and Vernon Bailey, who had overseen its transition to predator control, attended these meetings and commanded great respect from their fellows. Merriam, legendary as the mind behind the idea of altitude-based life zones around the world, was the mammalogy society's first president.

Suggestions for reform centering on Grinnell's idea began at the society's annual conference in 1924. At first the scientists were polite to the point of deference in their questioning of the bureau. By this stage everyone knew the bureau had reduced wolves, even in the national parks, to a shadowy handful of animals and that its hunters had put out more than 3 million poison bait stations for coyotes, killing untold thousands of nontarget birds and animals in the process. Scientists began their attempt to rein in the runaway program with papers emphasizing what a permanent step extinction was and the benefits predators conferred on "more valuable" ungulates by "removing weak and sickly animals" infected with diseases like septicemia or lumpy jaw.

Charles C. Adams, a founder of the Ecological Society of America, took the argument gently to the next step with a talk titled "The Conservation of Predatory Mammals." It directly supported Grinnell's national park idea. In the national parks surely "there will be less need of predatory control," Adams thought. After all, midsize predators like coyotes, he argued, "materially aid in rodent control." Then he continued, "Without question our National Parks should be one of our main sanctuaries for predacious mammals." But if the parks were expected "to make the predacious fauna safe," they were going to have to be larger, and there were going to have to be more of them.

The bureau's Major Edward (or E. A.) Goldman—whose family had migrated from Pennsylvania to California in the nineteenth century and along the way had changed their name from Goltman to Goldman—responded to these first expressions of discontent from the scientific community. A highly accomplished field naturalist who had helped do foundational work in the natural history of Mexico, Goldman was becoming a wolf specialist, and he was now on his way to becoming a major figure in the coyote's story. Goldman doubled down in his denial of a role for predators in America. Although he was "loath to contemplate the destruction of any species," Goldman told the audience, surely as biologists they must know that the bureau had no cho·ce but to "decide against such predatory animals as mountain lions, wolves, and coyotes." Stepping away from current bureau policies would alienate both hunters and the livestock industry. The criticism he was hearing clearly rankled Goldman. He felt himself grow warm as he thundered at the assembly, "Large predatory mammals, destructive to livestock and to game, no longer have a place in our advancing civilization." And that was that.

As far as some of the scientists were concerned, the bureau had thrown down the gauntlet. The society managed before its 1924 conference ended to create a committee to draw up a plan "for the preservation of predatory mammals" and for public-lands preserves to provide them a refuge. Obviously some in the scientific community were not going to back down. But at the Bureau of Biological Survey, those in the higher echelons realized that a line of sorts had been crossed. As it pondered the meeting of 1924, its public relations statements seemed to pull back ever so slightly. In 1925 the bureau's annual report even appeared to embrace the Grinnell idea, concluding, "Little objection can be raised to the continuance of a limited number of predatory animals in national parks and in wilderness areas remote from civilization," although the newfound tolerance the report implied conflicted seriously with the bureau's simultaneous assistance in wiping out predators as fast as possible in those very parks.

Goldman himself drew a different lesson from the 1924 conference, and in his fuming he penned an article the following year that in his mind settled the issue once and for all. In "The Predatory Mammal Problem and the Balance of Nature," Goldman minced no words. Ecologists who prattled on about predators and the balance of nature were conveniently forgetting that the arrival of people from Europe had changed everything about North America. The balance of nature might have been fine for Indians, but with white people on the scene, the balance of nature on the continent had been, as he put it, "violently overturned, never to be reestablished."

As a result, leading ecologists remained wary and unconvinced that the bureau was going to be reasonable. The growing rift between scientists and the bureau festered throughout the late 1920s, and it got worse when the mammalogists learned from the bureau's own figures that since 1924 government hunters had put out another 2,174,886 poisoned bait stations across the West. Now under the leadership of Stanley Young (who in public appearances began to imply, fraudulently, that he had a PhD), in 1929 the coyote-hunting division of the bureau gained a new name, Predatory Animal and Rodent Control (PARC). During its first full year, PARC hunters set out 181,887 bait stations in Colorado alone. True enough, in the wake of the scientific mutiny, the bureau's director in the late 1920s, Paul Redington, tried to downplay the word "extermination" in his public pronouncements. In another stab at political correctness, he got the Denver Eradication Methods Lab's name changed to the Control Methods Lab. But even he demonstrated telltale disbelief: "We face the opposition," he announced incredulously in a talk to the faithful, "of those who want to see the mountain lion, the wolf, the coyote, and the bobcat perpetuated as part of the wildlife of the country."

At a planning conference in Ogden, Utah, in 1929, the bureau finally decided to take a stand against the Grinnell idea of allowing predators a refuge in the national parks: "We cannot favor sanctuaries for the breeding of mountain lions, wolves, bobcat, and coyotes," the conference resolved. That same year, at the American Game Conference in December, chairman Aldo Leopold set out the evolving counterposition of the scientists with respect to predators. "No public agency" (guess which one) should control predators without substantial research first. Poisons should be an "emergency" control only. And "no predatory species should be exterminated over large areas."

At their 1930 annual meeting at the American Museum of Natural History in New York, the mammalogists organized a panel specifically on predator policy and invited Edward Goldman and Vernon Bailey of the bureau to express their views. By 1930 the bureau had for all purposes extirpated wolves from the Lower 48 and advised and assisted in erasing gray wolves from the crown jewel national parks, Yellowstone and Glacier. If the scientists expected rueful regret about this, they were in for a rude awakening. In New York Goldman let the audience know in no uncertain terms that extermination was still the game, and now it was the coyote's turn. His only qualification had to do with the science behind the policy, but secretly he thought studies then underway that would all but convict coyotes. So he put the matter to the panel this way: "It seems a reasonable forecast that additional studies will confirm the conclusion that the coyote is the archpredator of our time."

Yellowstone National Park was one of the laboratories where the bureau's E. A. Goldman believed science would convict coyotes of high crimes against nature. It was also one of the prime locales where 1920s biologists thought we ought to protect and finally study predators. The world's oldest national park has served as a setting to untangle natural relationships in America in almost every way imaginable for the past 150 years. So it's no surprise that if you yearn to understand the role coyotes played in the changing ecology of the Biological Survey's America, Yellowstone is one of the places to be. Coyotes and wolves, family cousins in a very old dog-eat-dog relationship, are encountering one another in several locations around North America now. But nowhere else provides quite the lessons as here for grasping coyotes as predators versus wolves as predators or how the gray wolf's presence and absence have influenced the coyote biography. That was true in the 1930s, and it still is.

The Yellowstone Wolf Project people, Doug Smith and Rick McIntyre in particular, are amenable to a visit to watch this ancient canine relationship renewed and revisited. So Yellowstone, the scene of so many bureau/biologist debates three-quarters of a century ago, is where you go.

You can drive the road that traverses the marvelous Lamar Valley, in the once ignored northeastern quadrant of the park, just about any time of year and expect to see wolves and coyotes. But my fiancé Sara and I pick September, the weekend of the autumn equinox, for one last stretch of perfect Indian summer days up on the 7,000-foot Yellowstone Plateau. Doug, a major player in the 1995 release of the first Canadian gray wolves in the park, sets us up with Rick to go into the field, but first he offers me some thoughts. You could say, and you would be right, that on the topic of wolves and their effects on the Yellowstone ecosystem, Doug Smith's thoughts are some of the savviest around.

Smith looks exactly as you'd imagine a Yellowstone Park wolf biologist to look. Tall, fit, and handsome, with a Sam Elliott mustache, Doug is a little gray and grizzled at fifty-two. He got his PhD from the University of Nevada, then worked with some of the most famous wolf people in the world, including David Mech, in the few remaining wild wolf outposts in the Lower 48, Isle Royale National Park in Michigan among them. He was part of the crew when the park and the US Fish and Wildlife Service formed the Yellowstone Gray Wolf Restoration Project in 1994. Now, twenty years later, he is its head biologist and project leader. Generous with his time and knowledge, he lays out for me the basic elements of a fascinating story.

"In general," he tells me, "across North America, where you have wolves, you don't have many coyotes." A westerner, Doug pronounces the word *KI-ohts*, emphasis on the first syllable. "And where there

The Archpredator of Our Time

Gray wolf tracks, Montana, 2009.
Courtesy Dan Flores.

are wolves, coyotes hunker down close to people, they get in close to towns, settlements, farms where they can use people as cover. They prefer people, who are more benign, to a wolf who's stalking you all the time." But Yellowstone is one of the places where wolves and coyotes hunt the same landscapes now. For the past two decades the park has given us a rare opportunity to observe how nature, disassembled with wolf extirpation in the 1920s, is stitching itself back together with the species' restoration.

The experience of the Yellowstone coyotes when the wolves returned after an absence of seventy years is now a famous park story. "Before wolves the coyotes were the big dogs on the block. Then we introduced gray wolves out of Canada to the park, and they went swaggering through places like the Lamar Valley, putting the fear of God in those coyotes. I forget exactly how many dead coyotes we documented over the first few years." Doug pauses. "But there was, you know, a real big spike in dead coyotes two, three years after wolf recovery, and 90 percent of them were at the elk carcasses that wolves killed. . . . I recall over one hundred dead coyotes the first two years." In that initial set of encounters, the elk carcass banquet wolves provided in places like the Lamar Valley was simply irresistible to local coyotes. They'd had the run of the valley for decades, so at first they'd darted boldly in to snatch a meal when the wolves were sated and meat drunk. On occasions it worked; most of the time, the coyote ended up dead.

Yet unlike at Isle Royale, where wolves entirely wiped out the small resident coyote population, the Yellowstone coyotes survived. "Now we hardly ever pick up a dead coyote killed by wolves," Doug adds. "I can tell you from walking around Yellowstone all the time, there are good coyote numbers out there and that suggests some type of coexistence now."

This brand-new wolf-coyote interaction is important to track, it seems to me, for what it tells us about the deep evolutionary history of both species. In the dimness of continental history, coyotes evolved to occupy a niche adjacent to and in conjunction with the one wolves occupied. The current scientific argument holds that coyotes acquired many of their behavioral traits—including their nervous wariness and the stunning intelligence that allows them to survive so well in our midst—by living in close association with their dangerous larger relative, the gray wolf. The return of the gray wolf, in particular, to Coyote America has come as a bit of future shock, and not just for those used to an America without wolves. It is future shock for the coyotes, too, which must have thought a world without gray wolves was coyote nirvana.

If the coyote is what it is not because of a comparatively recent relationship with humans but because of its ancient and evolutionary relationship with wolves, then understanding coyotes within the context of an intact wolf community seems crucial to knowing coyotes. If we want to understand why coyotes have been such a success in modern history, why they're hunting geese along the lakeside in Chicago and learning to wait for traffic to pass on interstate highways, we have to look directly at how they came by adaptations that

The Archpredator of Our Time   129

Observation of wolves and coyotes interacting with one another and with prey in Yellowstone National Park has become critical to biologists' and the public's understanding of how predation worked in ancient America.
Courtesy Dan Flores.

A pair of coyotes joins a wolf, a grizzly, and ravens over a carcass in Yellowstone Park.
Jim Peaco photo, 2013. Courtesy Wikimedia Commons.

made one of the most persecuted animals in America also its most wildly successful one.

Watching wolves and coyotes in Yellowstone's Lamar Valley is a bit like going to the premiere of a major new film in Los Angeles. Several dozen nattily dressed people, attired in the most recent offerings from Eddie Bauer and REI, their late-model Jeep and BMW and Mercedes SUVs parked nearby, are in a kind of receiving line, peering through their Swarovski spotting scopes at the celebrities, who consist primarily of the park's new wolf population. Fall in among this group and unless host Rick McIntyre points someone out—or you run into folks you know—you can end up unaware that you're rubbing shoulders with any manner of famous people from around the globe. Writers and journalists, well-known academics from far-flung universities, superintendents of other national parks, they're all among the throngs of people assembled at the valley turnouts for these events. So are a large number of people from small communities near Yellowstone who are obviously passionate about predators.

With amiable and well-connected park public coordinator Rick McIntyre as our guide, Sara and I are quickly immersed in a predator observation culture that has come to thrive in Yellowstone over the past two decades. Red-haired baby boomer Rick is a native New Englander, educated at the University of Massachusetts by old-school natural resource professors who never could have appreciated something like deliberately restoring gray wolves to a national park long lacking them. Rick introduces us to a congenial and generous group, all fans of charismatic predators, all with a fanaticism for meticulous observation, who clearly know individual animals by name or number and follow them through all seasons across the sprawling geography of the park.

In late September 2013, the sought-after movie stars are the members of the Junction Butte wolf pack. The Gray Wolf Restoration team released seven wolf packs from Canada into the park in 1995 and 1996, with the Crystal Creek pack of 1995 and the Druid pack of 1996 eventually battling it out for control of the Lamar

Valley. The Junction Butte pack consists this September of eleven wolves, three sisters and two brothers among them. Nine are gray; two are black. To avoid humanizing them, park personnel don't give either wolves or coyotes names, but the volunteer wolf watchers call the gray alpha female of this pack "Ragged Tail." The alpha male is a tall, thin gray wolf known affectionately as "Puff," a name bestowed on him in younger days when sarcoptic mange left him with mere patches of fur.

This first morning Rick has located the Junction Butte wolves using radiotelemetry equipment, and as we watch them string out across a sagebrush slope four miles from our overlook, Rick points out Puff, now fully recovered and "probably the fastest wolf" in the pack. "Even though he doesn't give the appearance of being a really strong wolf, he is. He's become maybe the best hunter in Yellowstone," Rick adds.

That comment becomes a prophecy. Later that day Puff almost single-handedly pulls down a young cow elk in the Lamar Valley, and for the rest of the afternoon the admiring roadside crowds train their Swarovskis on the scene and exclaim as if Paul Prudhomme were preparing a tableside meal on Jackson Square. We do the same, but as the pack devours the elk, I can't help noticing that there is not a coyote in sight.

Twenty-fours hours later, after members of the Junction Butte pack have eaten their fill, then trotted some six to eight miles west to rest and digest, we set up our spotting scope to glass the kill site from the day before. And now, with the wolves a safe distance away, we at once see coyotes where there had been none. Two of them, animals that appear, through our scopes, to be fully grown adults, are on the kill site, scavenging what remains. In the midst of an entourage of ravens and magpies, one of them is bracing itself and pulling up what seems, even at this distance, to be fairly large chunks of bone and flesh. The two coyotes work the carcass by turns for about fifteen minutes before a third coyote, what some biologists refer to as a young transient, or "floater," not attached to a pack, approaches from the river. At that point one of the original tandem dashes out to deflect this newcomer, and after a short pursuit, coyote number three retreats.

These coyotes are in the right geography to be descendants of a prewolf coyote pack here known as the Bison Peak Pack, one of eleven contiguous coyote packs once stacked neatly side by side, like eggs in a carton, up and down the Lamar Valley. We are looking at these coyotes across three-quarters of a mile of distance, and even through a 50X spotting scope, they appear much smaller than the wolves on this kill the day before. Otherwise their actions are similar, and indeed some of the tourists watching this morning seem to think they're observing gray wolves. Through atmospheric heat-wriggles magnified by the spotting scope, we can see one of the coyotes scent-rolling in what remains of the carcass, while the other—bracing all four of its feet, its tail curled under in an arc—manages to pull the carcass entirely off the ground and drags it backward and uphill toward the trees, sending ravens and magpies into spiraling, flapping, protesting flight. We are too far away to hear the cacophony of bird insults, but watching, it isn't too difficult to imagine the din.

"It's a pretty good bet those coyotes were there all the time, probably lurking back in the aspen groves yesterday," Rick offered. "It's taken till this morning for them to feel confident enough to approach that kill."

A Lamar Valley coyote's simple act of pulling the remains of a wolf kill uphill and into a forest rests on 20,000 years or more of competition between coyotes and gray wolves. As predators of pursuit, coyotes became one of the fastest animals in the world, slower than cheetahs or pronghorns, to be sure, but capable of speeds up to forty-three miles per hour. Only seven or eight animals in the world are faster. Nonetheless, on an open plain, wolves can run down a coyote. Among the morphological changes and learned behaviors coyotes bring to bear on their relationship with their wolf cousins, one is to engage with them, when possible, in either forested or hilly settings, where the smaller, quicker coyote can dodge and weave and outmaneuver a wolf or can lead a pursuing wolf downhill, then quickly swap directions and escape uphill while the heavier wolf spins out on the turn.

One scientist associated with the study of Yellowstone coyotes is Bob Crabtree, who in 1989 landed a National Park Service grant to examine the park's coyotes on the eve of wolf recovery. Crabtree's PhD is from the University of Idaho, where he studied a similarly protected coyote population on the Hanford Nuclear Site in Washington State. He's also an activist on the board of the San Francisco–based Project Coyote and has sparred in print with the Division of Wildlife Services, the successor to the Bureau of Biological Survey's PARC. One interviewer described him as a scientist who couldn't string together seven words without including an expletive. I liked the guy already.

Crabtree has used his work on those rare coyotes protected from both wolves and humans as a kind of baseline, and he argues that when coyote populations are unmolested—the situation at Hanford, and Yellowstone from the late 1930s until wolf reintroduction in the mid-1990s—their numbers stabilize at the carrying capacity of the local landscape. Crabtree's work indicates that before wolves returned, the Lamar Valley had a coyote population of eighty animals, most of them members of one of the eleven packs there. In this situation, even in a game-rich national park, coyote litters were slightly small, averaging 5.4 pups, with only 1.5 pups per litter surviving into autumn. Coyote populations thus were stable across decades. It's what could have happened in 1920s and 1930s America once the bureau wiped out wolves. But because of bureau and ag community intransigence about coyotes, of course it didn't.

And what happened when wolves returned and at once began to harass and kill coyotes? According to Crabtree, within three years the Lamar Valley coyote population dropped from eighty to thirty-six. Most of those coyotes were killed outright by the wolves, the overwhelming majority in the battles around carcasses that Doug Smith had mentioned. Sometimes, as at the Lamar site still known to the biologists as "Dead Puppy Hill," wolves even excavated coyote dens and killed the pups inside. No wonder the coyotes we watched in the valley waited until the Junction Butte wolf pack was eight miles away before visiting that kill site.

Ultimately, whether it's persecution by gray wolves or by humans, harassment and killing of coyotes triggers the same set of responses, adaptations hardwired by evolution into coyote genes. As the coyote sees the world, allow it to live out its life without its neck on a guillotine, and it will rear a sufficient number of pups to reach the carrying capacity of its territory and no more. But it's had to survive less benign circumstances for many thousands of years. Hound it and mark it for extermination, and not only will its biology defeat you every time, but it will colonize into new settings where you never in your wildest dreams expected to confront yellow eyes peering out of the twilight.

If anyone in the scientific community doubted Major E. A. Goldman's sincerity about the coyote being the "archpredator" of 1930s America, the Animal Damage Control Act of 1931 disabused him or her of any illusion that the bureau would renounce extermination. The act originated in 1928 in a Texas congressman's plan to wipe out coyotes once and for all. Once proposed the idea gained momentum and precision: Congress would appropriate $1 million a year for ten years "to promulgate the best methods of eradicat[ing], suppress[ing], or bringing under control" primarily coyotes but also a suite of other "varmints" the United States regarded as unworthy wild citizens of the republic.

Almost 150 scientists from some of the most distinguished universities and research facilities in the country at once signed a petition protesting the Animal Damage Control Bill, arguing among other things that wholesale attempts to eradicate coyotes would produce enormous collateral damage among innocent animals. Even the bureau's now elderly founder, C. Hart Merriam, came out publicly against doubled-down poisoning. But extermination advocate Scott

Leavitt of Montana agreed to serve as the chief sponsor of the bill in Congress, and Edward Taylor of Colorado—soon to become famous as the architect of the Taylor Grazing Act, which ended homesteading in the United States—was his wingman in the effort to "clean out this scourge."

Despite a heroic battle waged by the mammalogists in testimony, newspapers, and magazines like *Outdoor Life*, Congress passed the bill, and President Herbert Hoover signed the Animal Damage Control Act into law on March 2, 1931. In its claimed goals of protecting livestock, saving valuable game animals, and suppressing diseases like rabies, the act specifically designated the "national forests and other areas of the public domain," along with state lands and private holdings, as staging grounds for the eradication or "control" of predators. Notably, the bill did not mention the national parks. Indeed, Joseph Grinnell's proposal that predators remain unmolested in the parks would become the single victory the scientists realized from taking on the bureau's policy of extermination. And even that concession would wait until 1935.

As Major Goldman had promised the year before, this time the bureau pinned the species-cleansing bull's-eye directly on America's junior wolf. Gray wolves were gone; Mexican, red, and eastern wolves had dwindled to a few straggling, shadowy remnants. Now, with $1 million a year to spend on the prospect, the bureau deemed it señor coyote's turn for shock and awe.

But it turned out that the bureau still faced hurdles. Goldman's reference during the 1930 panel in New York to "additional studies" pointed directly at the work of two mammal ecologists who would become legendary figures in twentieth-century American conservation. At the time, however, they were, in effect, working for the man, with the stated goal of buttressing the case against the scientific critics who were fighting the bureau's predator policies. The Minnesota brothers, Olaus and Adolph Murie, were born ten years apart in the late 1800s. Both had taken advanced degrees in the new field of wildlife management at the University of Michigan, and both began their careers as government biologists. The older brother, Olaus, in an odd quirk of fate, had gone to work for the Bureau of Biological Survey in 1920 and quickly impressed his superiors with a landmark study on elk in Jackson Hole, which he completed in 1927.

The Murie brothers would become famous for a shared conviction: that scientists must above all be ethical. When Olaus's superior, E. A. Goldman, approached him with the proposal that he study Jackson Hole's elk predators, coyotes, Goldman expected Murie's research to show that coyotes were archpredators whose depredations on stock and game animals had grown so egregious that the species deserved a death sentence. Remarkably, younger brother, Adolph, working for the Wildlife Division of the National Park Service, was taking on a similar project in Yellowstone at almost the same time. Olaus Murie's *Food Habits of the Coyote in Jackson Hole, Wyoming* appeared in print in 1935, followed five years later by Adolph's *Ecology of the Coyote in the Yellowstone*. The bureau would coldly ignore both volumes. Yet they turned out to be extremely important studies in the scientific firestorm that never stopped singeing people around the Animal Damage Control Act from the 1930s on.

The problem all along for both camps was that no one from either side had good, untainted evidence about coyote natural history or food habits. The bureau had accounts from ranchers and sheepmen about coyote predation on calves and sheep, and its hunters turned in reports on coyote stomach contents from animals they poisoned and trapped. But the former were hardly unbiased; they attributed kills to coyotes whenever they saw them on carcasses, even though the original cause of death was often unclear. As for the bureau's field men, they were entirely untrained in scientific techniques of analysis. Yet, while Kaibab and Kern County seemed to support intuitions about the balance of nature, in truth the scientists had no good studies on the diets of coyotes either.

The two Murie brothers aimed to correct that. Olaus, for his part, conducted a classic "stomach contents/scat study" in a straightforward effort to determine what his coyote subjects in Jackson Hole ate. Younger brother Adolph, trained a decade later, was more up-to-date and more thorough. His *Ecology of the Coyote in the Yellowstone* was an actual ecological study, analyzing both predator and its various prey animals, along with the influences of weather, disease, and even history. Not appearing in print until 1940, five years after the national parks were excluded from the bureau's extermination campaign, Adolph's study would even take on some of the bureau's and its supporters' most treasured (but untested) myths about coyotes.

Reading these two studies today, it is clear why Olaus's Jackson Hole study infuriated his bureau superiors, who without any evidence whatsoever had succeeded in painting coyotes as a bane of nature and civilization. For four years, from 1931 until 1935, Olaus collected data, much of it from the Teton Game Reserve, with its large elk population. Yet by the end of his work, he had to conclude that mice, gophers, and hares were the chief prey of Jackson Hole's coyotes. Rather than archpredators of game animals, coyotes turned out to be omnivorous generalists. Murie found that the 1,629 dietary items he analyzed represented twenty-eight different mammals, ten birds (including their eggs), and, in a sampling he admitted was probably weak, nine items that were either fish, insects, or plant matter. An archpredator in the vicinity of an elk preserve ought to have been meat-drunk with elk, and Murie did find elk in coyote scat and stomach contents. But he concluded from observation that virtually all of it was scavenged carrion. Ethically, he had to report that coyotes appeared to be "unimportant" in elk predation.

Nor could Murie tell his bureau superiors that coyotes were major predators of mule deer, mountain sheep, or antelope. From the coyote's point of view, everything it ate was beneficial, of course. But Murie's analysis indicated that, all things considered, 70.3 percent of a coyote's food sources produced a net benefit for humans as well. Another 18 percent of the junior wolf's wide-ranging diet had a neutral effect on human endeavors. So much for Goldman's "archpredator of our time."

Adolph's Yellowstone work was far more sophisticated, but it yielded similar conclusions unwelcome by government bureaus that had dedicated themselves to wiping out a coyote "scourge." The two years of fieldwork that went into *Ecology of the Coyote in*

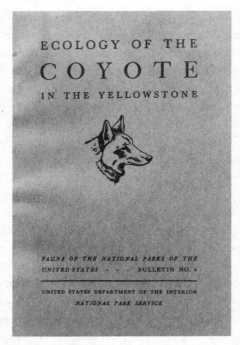

Biologist Adolph Murie's breakthrough study exonerating coyotes as stock-and-game predators appeared soon after Congress authorized a program to exterminate them.
Courtesy Dan Flores.

the Yellowstone took place in a distinctive context. Twenty years after Joseph Grinnell had proposed the idea in *Science*, predators now had refuges—even from PARC hunters—in America's national parks. As the *Journal of Mammalogy* had announced, "Predatory animals are to be considered an integral part of the wild life protected within national parks, and no widespread campaigns of destruction are to be countenanced." But many were the dire predictions of doom *that* produced. Retired Park Service director Horace Albright had done all he could to squash any sympathy for coyotes. In fact, sentiment in favor of restoring coyote control in the parks had gotten Murie's work in Yellowstone approved in the first place. His superiors had thought his study would result in a return to poisoning and trapping.

So Adolph Murie knew he had a keenly interested and suspicious audience no matter what he found. Like a twenty-first-century climate scientist, he knew he was conducting critical research in a highly politicized atmosphere. He responded in much the same way as today's climatologists. He was exceedingly careful.

Murie began not in the field but in the Yellowstone archives, trying to develop an understanding of the history of coyotes on the Yellowstone Plateau. Aldo Leopold said in 1936 that he had found no coyotes in the mountains in Mexico, and there was a general sense all along the Rockies that coyotes had arrived recently in the high country. So, were coyotes in the mountains originally? The historical record seemed clear that coyotes had always been more numerous on the Great Plains, but Murie found ample indications of them high in the Yellowstone country too, at least as far back as the writings of trapper Osborne Russell in the 1830s. Indeed, contrary to Major Goldman's insistence that the arrival of white Americans had destroyed the balance of nature, Murie concluded from his research that the coyote's relationship "to the rest of the fauna is today similar to what it was formerly."

Murie also discovered that as early as the 1890s, tourists in the park had become interested in and favorably disposed toward the coyotes they saw. But as he wrote in his foreword, park superintendents had always insisted that Yellowstone was supposed to showcase game animals and that controlling coyotes was essential to preserve deer, antelope, and bighorn sheep for those same tourists. That opinion had continued even after control efforts ended in 1935, accompanied by dire warnings that in the absence of human control, the coyote population would certainly skyrocket with disastrous results. Yet, Murie found that four years after the park had ended coyote control, the coyote population in Yellowstone remained at about the 1935 level. Contrary to predictions, there simply had been no explosion of coyotes.

Since the park's resident coyotes had not swelled in numbers, it followed that they had also not spilled across park boundaries to wreak havoc outside the park, as so many had warned. In the adjacent Absaroka National Forest, rangers had estimated the coyote population in 1935 at nine hundred animals. After coyote control ended in Yellowstone, the national forest population actually fell, to 781 animals. So much for yet another hand-wringing prediction. Why, in any case, would an animal as smart as a coyote want to leave protection and plentiful food in a national park for the dangers of trappers and poison bait outside its boundaries? As Murie argued, the vast majority of coyotes born in Yellowstone would very likely die right there.

In the wake of the Animal Damage Control Act, Murie knew the crux of his investigation was to discover whether coyotes were truly archpredators of valuable game animals. Like his older brother, Olaus, he did this primarily by examining stomach contents and scat, in which his analysis ultimately identified nearly 9,000 food items. And like his brother, Adolph carefully enumerated the mammals, birds, invertebrates, and vegetables these represented. Some of the items that had passed through his coyote subjects definitely fell outside those categories. A piece of window curtain, a strip of rubber, a paint-covered rag, a gunny sack, a piece of towel, part of a shirt, and eight inches of rope made Mark Twain's *Roughing It* description from seventy years before sound like a masterpiece prophecy of coyote natural history.

But—and this, after all, was the whole point of the study—did coyotes prey harmfully on prized game animals? In case after case, Murie had to conclude that the answer was no. Primarily coyotes ate mice and gophers (55 percent of their diet) and carrion from large mammals (in Yellowstone about 17 percent of their diet). Grasshoppers and crickets accounted for nearly 10 percent of what they ate. Otherwise their menu was a wild smorgasbord, including the odd length of rope. Did they prey on elk? No, not even—except in anomalous circumstances—on elk calves. "All available data indicate that the coyote is a minor factor in the status of elk." What about mule deer, whose very survival, some had argued, depended on vigorous coyote control? In bad winters of crusted snow, coyotes sometimes took the weakest fawns, running them down slopes and catching them at the bottom. Otherwise, "there was no evidence that the coyotes molested any deer except fawns." Antelope, those beautiful, striped survivors from ancient America? Coyotes had been thought a threat to their very existence by people who blamed every antelope death on coyotes. Murie found some fawn predation but not at any problematic level and concluded, "The coyote is not at the present time adversely affecting the antelope." Bighorn sheep? "Coyote predation is at most an unimportant mortality factor."

Adolph Murie no doubt went to press in 1940 with the kind of expectations that wary climate scientists anticipate in our own time. For nearly a decade a powerful government agency, seconded by the Park Service he worked for, had moved heaven and earth for the extinction of coyotes in part because they were supposedly "archpredators" of valuable game animals. Now, in the first comprehensive study of the matter, a government-employed scientist said flatly, without any equivocation, "The facts show that in the case of elk [coyote predation] is negligible, and that no appreciable inroads on the populations of deer, antelope, and bighorn are taking place."

The Murie brothers did many other things across their illustrious careers, but neither ever lost a fascination with coyotes. Through the 1930s and 1940s Olaus continued to assemble data on coyote diets in places like Montana and British Columbia. He illustrated articles for J. Frank Dobie while Dobie was working on his folkloric book *The Voice of the Coyote*, and he collected newspaper articles on anticoyote sentiment among stockmen and hunters. He and brother Adolph even raised a female coyote pup as a pet.

While still working for the bureau, Olaus Murie went so far as to evaluate "the factions interested in [the] coyote question" for his superiors. He concluded that there was an emerging group he called the "Nature Lovers," and to the evident disgust of his superiors he argued that this group might actually represent the future and a state of enlightenment with respect to predators. "I firmly believe that it is working against the best interests of humanity to . . . . ridicule those who see beauty in a coyote's howl."

Eventually Murie came to think that he and other scientists of the period had misunderstood their own motivations during the 1930s. Before a public they knew the bureau had brainwashed into hating predators, the scientists had opposed poisoning because of collateral damage, because of all the innocent animals that ended up dying along with coyotes in the bureau's war of extermination. But in a famous letter he would write in 1952, Murie asserted that, in truth, "concern for the coyote itself" had turned so many scientists against the bureau and its coyote policy.

Concern for the coyote itself. Murie knew that once, in his Yellowstone research, his brother had stood rapt, watching a coyote trot along a trail with a sprig of sagebrush in its mouth. At repeated intervals it had tossed the sprig joyously into the air, caught it, then trotted on. Why had so many in the bureau, without any science to back them up, so hated an animal that took that kind of pleasure in being alive in the world? Why had they encouraged hatred for coyotes among the public? It was not an attitude or a culture Olaus cared to be associated with. He ended up leaving the bureau for the Wilderness Society.

The several score American mammalogists who were by now heavily invested in the coyote question felt vindicated by the fieldwork of the Murie brothers, but they knew that exonerating coyotes as major predators of game animals was only part of the task. With wolves all but gone in the Lower 48 and mountain lions rapidly dwindling, hunters would never have to be arm-wrestled into believing coyotes were rivals to filling their elk and deer tags. They would ignore the Murie brothers' research. It was the livestock industry, especially sheepmen but even some cattlemen, that kept the pressure on Congress and the bureau's PARC hunters to continue the coyote campaign. And as they had long done, stockmen's associations and state legislatures contributed external money when New Deal contingencies left the bureau a little short of those promised million-dollar-a-year appropriations.

Some incredulous coyote from hundreds of years before had been the first of its kind to marvel at what a lob across the plate the domestic sheep was. By the 1930s generations of western coyotes had never gotten over marveling, so sheepmen, in particular, were shrill with some justification. The country's coyote population was likely growing in the 1930s for two reasons: carpet-bombing lethal control, which kept coyotes constantly colonizing and rebuilding their populations, and what ecologists today call "mesopredator release," which resulted when wolves faded away. The mouflon variety the sheep industry had imported to America descended from a wild sheep hunted in the Old World by the coyote's distant cousin, the golden jackal. Rendered incapable of resisting shepherds—and therefore predators—through 8,000 years of domestication, sheep in that helpless condition had then been plopped into the heart of coyote habitat in the West. With millions of normal prey—rodents and prairie dogs—vanishing before the bureau's poisoning onslaught, coyotes in sheep country did exactly what a coyote had evolved to do: they laid into such easy sitting ducks. And with the number of domestic sheep in the United States reaching 56 million during World War II, coyotes must have been drunk with their good fortune.

But the bureau's all-out war against coyotes repeatedly disrupted their normal social behavior, and for sheepmen the results were not good. The poisoning of older alpha males and females, gassing and strangling of pups, and harassing and winnowing of coyote social groups resulted in inexperienced beta animals assuming alpha status and breeding. That had the potential to send coyote social life into a loop of fractured, abnormal behavior not unlike what history has recorded for the Indian populations of the Americas in the wake of contact with Europeans. Waves of virgin-soil epidemics caused by Old World pathogens killed millions, among them elders and priests responsible for conveying cultural lore to younger generations. Hard-won knowledge of how to exist in the world vanished overnight.

Something like that seems to have happened with some coyotes. Young coyotes were surviving without proper cultural training, and with the pressure of raising litters, they attacked livestock, especially marks as easy as sheep. Coyotes may not have been archpredators of the natural world, but domestic sheep were low-hanging fruit. Killing them required little training in the hunt.

Whether because they are retiring introverts or for some other reason, scientists have never been particularly good at public relations. Judging from the climate debate, they still aren't. In the 1930s and 1940s, they were getting their hats handed to them by the bureau, which was clearly winning the war for the hearts and minds of the public. In the mid-1930s, newspapers around the country, among them even the *Washington Post*, ran an illustrated, canned bureau article that, in the age of John Dillinger and Pretty Boy Floyd, offered up coyotes and other predators as the "gangsters of the animal kingdom" and characterized bureau hunters as the heroic G-men who would protect society, "man and beast, against the animal underworld." It was a clever set piece, and it worked. Despite the Murie brothers' findings, in the public mind coyotes deserved the same fate as Bonnie and Clyde.

In 1935 Ira Gabrielson replaced Ding Darling as director of the bureau. Gabrielson (who conservationist Rosalie Edge said could make poisoning a coyote sound like a beautiful experience for poisoner and coyote alike) was still in charge in 1940 when the bureau, transferred from Agriculture to the Department of Interior, became the Fish and Wildlife Service, the name it still carries. PARC made the transition with its name unchanged. Gabrielson would hold his position for eleven years, during a time when rumors of war, then real war on a scale never before seen in human history, absorbed the country's attention. All the same, the coyote war at home continued, and one of Gabrielson's field men saw all of this as part of a whole:

"I hope I have three celebrations coming—when we whip Hitler and Hirohito and when we kill that damn coyote."

So throughout World War II, and partially in response to the huge shot in the arm the war effort gave to widespread technological innovation, the smaller, less heralded campaign against coyotes went on. And on and on. In 1946, the year Gabrielson retired, two events of significance in the effort to exterminate coyotes from North America occurred almost back-to-back. In September, E. A. Goldman, architect of the phrase "archpredator of our time," passed away from a stroke, dying only a few months after retiring and while serving as president of the American Society of Mammalogists. Then, in a kind of tangible epitaph to Goldman, in December the Fish and Wildlife Service approved the use of sodium fluoroacetate and thallium sulfate against coyotes.

That final decade for wild coyotes in America, promised by the bureau back in 1931 if only Congress would pass the Animal Damage Control Act, had stretched to fifteen years. And coyotes were still on the scene. More than that, coyotes had reached the Yukon and Alaska and were showing up in one state after another in the Midwest, the South, and the East. Despite a focused and single-minded campaign against them unlike anything in American history, coyotes were still out there, now loping casually along boulevards, glancing back defiantly, for some inexplicable reason impossible to eradicate.

Hence the two new poisons. Strychnine had killed hundreds of thousands of coyotes in its century of use, but obviously it had fallen short of the task. Coyotes were especially bright and observant, and strychnine acted so quickly that other coyotes in the vicinity of a poisoning victim learned to be wary of the baits and their associated smell. Thallium sulfate had been around since the 1920s, mostly used against rats, but tests on coyotes at Denver's Wildlife Research Laboratory—another new name for the old Eradication Methods Lab—showed it had real advantages. It not only was odorless but, happily enough, killed coyotes slowly. Rather than producing the thrashing, struck-by-lightning reaction caused by strychnine, the new poison took days to kill. Coyotes that fed on a carcass baited with thallium sulfate went blind. They lost the pads on their feet. Their pelage dropped off in tufts. Naked coyotes poisoned by thallium sulfate were sometimes found huddled together, freezing and blind but not yet dead. But with the relationship between cause and result obscured, coyotes did not become bait-shy.

The second poison was even more effective. Sodium fluoroacetate occurs naturally in the Australian "poison pea" plant family but was synthesized into a commercial poison during World War II. The poison acquired the name Compound 1080 because it took the Wildlife Research Lab that many tries to perfect it. It was cheap, easy to handle, and simple to use. Injected into a bait animal like horse—a lethal coyote dose was 1.6 grams per one hundred pounds of horse—it would poison every molecule of flesh. Coyotes feeding on such a bait animal showed no symptoms for up to an hour, again long enough to confuse other coyotes about the cause of death. But there was no surviving 1080. Within an hour or two, a poisoned coyote would be seized by grotesque convulsions. It would utter piteous howls and bizarre vocalizations. Then it would run uncontrollably until it dropped.

The World War II era had produced an explosion of knowledge about chemicals and poisons, and along with these two, the Denver lab promoted yet a third, administered by a device the Fish and Wildlife Service, to its eternal discredit, called the "humane coyote-getter." Originally the device was a brass .38-caliber casing inserted into the earth and capped by the kind of scented cloths that Adolph Murie had found in the stomachs of coyotes in Yellowstone. If unable to resist these cloth offerings, which were suddenly springing up like mushrooms, the coyote ingested a tablet that exploded into a sodium cyanide mist in its mouth. What was "humane" about this? Cyanide didn't cause blindness or hair or pad loss. It left pretty corpses.

Everyone, from PARC hunters to outside observers, believed the trio of new poisons would finally accomplish what the old bureau had promised back in 1931. World War II had killed millions. Now, at least, the offspring of the science it had engendered promised to erase coyotes from the continent of their origin. No American animal had ever been the target of this kind of viciousness. In 1952 Olaus Murie, writing a review of the old bureau stalwart Stanley Young's *The Clever Coyote* for the Wilderness Society's magazine, adopted the position that the bureau's hatred had finally won, that species cleansing was already a fait accompli for the coyote. "Many who had formerly taken for granted the presence of Señor Coyote and his song, without much thought," he wrote, "now miss him, now that he is gone."

Despite the bureau's obsessively tabulated official kill of 1,884,897 coyotes from 1915 to 1947, rumors of the coyote's final song were exaggerated. But how and why? The poison trifecta of the postwar years does seem to have extirpated coyotes, at least locally, in many regions of the country beginning in the late 1940s. By 1957 PARC's budget from Congress had almost doubled, to $1.76 million. Added revenue from states and livestock associations brought the figure up to an astonishing $4.5 million. From 1945 to 1971 the federal coyote killing program would collect the carcasses of a staggering 3.6 million dead coyotes, although because of the way the new poisons killed, many in PARC believed they destroyed another 3 million coyotes in those years that were never found. And still coyotes persisted, although backed by this kind of money, the new poisons roughhoused them.

One place that experienced a regional near-collapse of coyotes may provide us with an answer about what factors were critical in keeping coyotes from going under when faced with chemical warfare on this scale. The Texas Hill Country, low juniper-covered hills west of Austin and San Antonio devoted to sheep and goats since Spain owned the territory, experienced the most intensive poison and trapping carpet-bombing campaign against wolves and coyotes of any locale in America. One researcher called the relentless forty-year program there "a massive human effort using all the tools and techniques which could be brought to bear." First wolves, then coyotes practically ceased to exist in the Hill Country. Why? The answer may lie in the land ownership history of Texas, which had entered the

Union in the 1800s with so many debts that the government allowed the new state to retain title to its lands to pay them off. This it did by selling off its landscape wholesale. The result was a giant state with no public lands and only one national park (Big Bend), located hundreds of miles from the Hill Country, that could serve as a refuge. Yet the struggle for survival of Texas Hill Country's wolves and coyotes would become one of the amazing wildlife stories of the 1960s.

So Joseph Grinnell and the scientists had managed a win that helped coyotes after all. That one concession, protection of predators inside the national parks, was certainly key in helping to save coyotes at mid-century. In Yellowstone, Glacier, Rocky Mountain, Zion, Bryce, Canyonlands, and the Grand Canyon, coyotes survived unmolested.

But national parks and scientist saviors alone do not explain the coyote's persistence. In the next two decades a pair of new studies would demonstrate a fuller truth: all along, coyotes had been saving themselves. Biologist Fred Knowlton would finally untangle the behavior of coyotes under assault, and a computer simulation by biologist Guy Connolly using Knowlton's insights—Connolly titled the resulting article "The Effects of Control on Coyote Populations"—would produce an almost mind-bending revelation.

With the species under siege from efforts to wipe it out, Knowlton and Connolly had discovered, the coyote's evolutionary colonizing mechanisms kicked in. With beta females breeding, fission-fusion in high gear, larger litters, and more surviving pups, even reducing the total population of coyotes in a given area by 70 percent—not just once but year after year after year—produced no appreciable effect on coyote population density! Even at a 75 percent annual reduction, it would take half a century to eliminate a population. And once killing stopped, within three to five years in-migration and coyote cues about carrying capacity would return population numbers to where they'd been when 75 percent control began!

Sisyphus kept pushing his rock up the hill, and Coyote kept rolling it right back down again. Short of carpeting the continent with our new nuclear weapons, nothing was going to get rid of coyotes.

Were we really willing to continue this level of cruelty in the face of unending failure?

———

As with coyotes, as with most mammals, as with us. Epigenetics determine who we are; that mysterious interplay between our hardwired genetic selves and the experiences we have, which turn off some genes and dial up the gain on others, shapes the beings we become. The age of ecology looming in front of us in the 1960s was one of those mass epigenetic experiences that would change America. Like so many millions of others, I became someone different then, transformed by my experiences, and I suppose open to transformation. And through what seemed yet another inexplicable turn in the coyote story, the period provided many of us with a new, very twentieth-century experience with the animal—that ambiguous yet endlessly intriguing canine—that stood as such an emblem of wild America.

Those ever-trotting, golden-eyed creatures of vertical red canyons and sere deserts were about to become far more visible in the culture and in our living rooms. I, for one, would be enthralled.

CHAPTER 5

Morning in America

In the spring of 1961, the year the weekly television show *Walt Disney Presents* (it would evolve into *Walt Disney's Wonderful World of Color* later that year) decided to position itself dead solid in the middle of the growing cultural controversy in America over coyote control, the show's producers had me squarely in their sights as audience demographic. In 1961, I was twelve years old and an undying disciple of Disney. I never missed an episode. I'd seen his 1942 anti-hunting film, *Bambi*, a year or two earlier, and I'd soon enough take in *Lady and the Tramp* at a local theater and nurse a grudge against dog pounds and dogcatchers from then on. So at an impressionable age, in a decade on the precipice of the age of ecology and the beginnings of the environmental movement, I was all in when revered Uncle Walt stepped into the nation's coyote debate with his hour-long animated film *The Coyote's Lament*.

Disney himself, amiable as always, introduced his initial film on coyotes (there would be five more), beginning by quoting from Mark Twain's description of the animal in *Roughing It* to provide a reference point for what humans thought of the species. But *The Coyote's Lament*, he averred, would finally tell "the coyote's side of the story."

Over the next hour the film introduced millions of young viewers like me to three generations of a family of rustic, country-bumpkin coyotes (replete with rural Southern accents), including a pup who serves as the rapt audience for the accounts the older coyotes share. Those accounts mourn the changes wrought by humans and their dogs "when man came West" and chased off coyotes' "vittles," fenced in the prairie, and crowded coyotes out. Their natural food sources diminished, they'd had to turn to domesticated fare like chickens and sheep, incurring man's wrath. Scenes in *The Coyote's Lament* portray the war against coyotes remarkably realistically, with set-piece scenes on guns, bounties, and traps. In this coyotes' version of American history, man "took our domain. . . . [W]e were here before you." Then, their perspective of the conquest of the West told, the pack gathers atop a mesa to howl the film's recurring theme song, "You Made Our Lives a Misery," one line of which goes (I no doubt gritted my teeth in sympathy), "We don't want to be your durn pet!"

"We coyotes do lots of good," one of Disney's coyotes tells his television audience, before channeling the essence of the new scientific consensus about coyotes: "We're what's known as the balance of nature." At the end of *The Coyote's Lament*, the coyotes' final message to the audience, delivered during a period when federal poisoners, bounty hunters, and state trappers were killing between 250,000 and 300,000 coyotes a year in the United States, is simply this: "When the time comes when you can't hear the song of the coyote, the West is going to seem a mighty dull place."

By the time the credits rolled, I and millions like me were avowed coyote aficionados, the first generation in American history to have its sensibilities shaped by nature programming on television.

It's a bumper sticker cliché to think of the 1950s as a simpler time, the "morning in America" that Ronald Reagan would invoke three decades later, but no decade in America's past has ever been simple. Compared to the ear-splitting thunderclap of the revolutionary

1960s that loomed ahead, the 1950s almost seemed like a collective inhalation. For more than a decade 200 million people seemed to be holding their breath in anticipation. Perhaps the largest generation gap in American life occurred between those who came of age in the 1950s and those whose worldviews were shaped by the 1960s. I have two older siblings who both were teenagers in the 1950s. The cultural rift between them and our parents, themselves children of the 1920s, was far narrower than the grand gulf of viewpoints that separated my siblings and me. The 1960s did that to many millions of Americans.

Scientists of the 1950s and 1960s, especially specialists in ecological relationships like predation and, most particularly, those interested in the effects their evolving fields had on public policy, already possessed a nuanced idea about America's native wild canids. They had a new appreciation of the role of predators in the natural world that had been growing in sophistication over the previous three decades. But for ordinary citizens, who read no scientific papers or journals and only barely knew that coyotes were the special target of a poisoning campaign paid for with their tax dollars, the fate of coyotes wasn't even a blip on the radar or the TV screen. Westerners certainly knew about coyotes, although often only as a category of creatures classified by state game laws as "nongame varmints"—legally killed by various methods at any time of the year, with none of the restrictions that hunting seasons or bag limits imposed on animals considered "game," which, of course, is where most wildlife money went. If coyotes registered any recognition at all among southerners and easterners, they did so merely as characters in western literature or films. Hardly anyone, even the scientists, realized at the time that coyotes were slipping quietly through places like Louisiana and upstate New York, year by year trotting toward the Atlantic coast and already experimenting with urban life in cities like Los Angeles, with Denver and Chicago on their radar.

In 1949 the star biologist of the first half of the twentieth century published his most important book, which would spread his fame far beyond the scientific community. Aldo Leopold had been writing both science journal and popular articles for decades. Since he penned his 1917 essay "The Varmint Question" praising "the excellent work" of the Biological Survey, his views about predators had evolved more than any other scientists'. From at least 1940 Leopold had understood that to have an impact beyond their profession, at least a few wildlife scientists would need "to contribute to art and literature." He may have been thinking about his own 1939 *Journal of Forestry* essay, "A Biotic View of Land," as it played a critical role in the thinking that went into his eventual masterpiece. However he came to its ideas, Leopold's 1949 book, *A Sand County Almanac*, became both a best seller and a crucial philosophical foundation for the ecology movement that was about to sweep America as part of the countercultural revolution of the 1960s.

Books and film are equally capable of rearranging the furniture in our heads, books perhaps the more so because the experience of reading is so private and provides such opportunity for pause, for deliberate consideration, for mental testing. For those who read and deliberated on it in the 1950s and 1960s, *A Sand County Almanac* was an absolute game changer. In gorgeous, poetic passages and vividly rendered scenes, Leopold introduced postwar America to the insights of a full career in the ecological sciences. For many his insights constituted near-epiphanies. Leopold was the first scientist willing to promote an ecological philosophy of living. He called it "the Land Ethic," and it included his "Golden Rule of Ecology": "A thing is right when it tends to preserve the integrity, stability, and beauty of the biotic community. It is wrong when it tends otherwise." Leopold did not say that an act was right when it tended to preserve humanity or economies, the easy position of a self-absorbed species. Instead he called on readers to think of the innate rights—among them the simple right to exist—of other creatures, an idea his followers subsequently called "biocentrism."

The essay in *A Sand County Almanac* that ultimately provided the most memorable scene in the book, however, detailed Leopold's own story of personal conversion and redemption. "Thinking Like a Mountain" was not merely an accessible statement, written for a popular audience, of Leopold's maturing view of predators. Over the next quarter century it became far more than that. For a generation of readers soon to be immersed in painful soul-searching about so many unexamined assumptions in American life, Leopold's story of shooting a wolf, watching the "green fire die in its eyes," and realizing eventually what a miscalculation he had made about predators, their role in the health of the "biotic community," and even himself, offered a culture-wide catharsis. We were wrong—I was wrong—his story said. But it's not too late for salvation.

A Sand County Almanac became a national touchstone, carried around in back pockets and backpacks much like Jack Kerouac's *On the Road* or Tom Wolfe's *The Electric Kool-Aid Acid Test*, literary preparation for a shift in the zeitgeist of the age.

On one level, the changes in America that produced the global age of ecology in the 1960s had to do with the maturation of advanced industrial societies after World War II and their ability to produce a modern standard of living that went beyond essential necessities and allowed ordinary citizens to begin to consider quality-of-life values. At the very moment that a different kind of life became possible, Americans also began to confront for the first time in history the self-evident consequences of industrial development without the scantiest regard for the environment. Policies pushing extermination for predators like coyotes and wolves seemed, to many, to spring from the same inattention toward capitalist excess that produced deadly smog in the Northeast and Los Angeles, oil spills off the Pacific and Gulf Coasts, and rivers so polluted with chemical waste that they caught fire. A "subversive science" in a capitalist economy, ecology lent the 1960s its symbolic word power, which morphed into the environmental movement.

Leopold's book meant that biocentrism and the much persecuted canine predator got snagged by the dragline of the 1960s, caught up in the roiling upheaval of antiwar politics, feminism, civil rights, Earth Day consciousness, and sex, drugs, and rock and roll. Every one of these would produce profound changes in modern life. Leopold's biocentrism arguably produced the most controversial legislation of the entire environmental revolution. It, too, changed the world.

If, in the 1960s, you thought at all about other species' innate right to exist, you eventually came to believe that the new synthetic poisons—many of them products of the chemical revolution that natural resource shortages had jump-started during World War II— very likely posed a threat to life of all kinds. The widespread radioactive fallout from US and Soviet nuclear bomb tests during the 1950s had spooked many people around the world anyway. The era gave us science fiction films like *Godzilla* and *Them*, where mutated animals were the monstrous result of the products of the new sciences of the age.

Chemical companies advertised poisons like Compound 1080 and DDT, cooked up by chemists prior to and during the war, as game changers in quality of life, better living through chemistry. Not everybody bought it, though. The furor surrounding another blockbuster book, science writer Rachel Carson's legendary *Silent Spring*, published in 1962, vividly demonstrated how mid-twentieth-century American attitudes about the natural world, including the fate of the coyote, were beginning to morph into values many 1950s conservatives found almost unrecognizable.

Silent Spring famously employed songbirds as its primary victims and metaphor for a world dramatically changed by the use of the insecticide DDT. But the book's warning about the profligate and unexamined use of insecticides to curb "undesirables" raised public consciousness about poisons generally. A biologist and successful writer (her earlier book *The Sea Around Us* won the National Book Award), Carson had actually spent much of her career as an employee of the US Fish and Wildlife Service, which had evolved out of the Bureau of Biological Survey. So Carson was close enough to agency divisions like Predatory Animal and Rodent Control (PARC) and Animal Damage Control to have an insider's sense of what was happening in the world around her.

Carson didn't invent her subject. She was preceded in her outrage by other American writers. In 1959 outdoor writer Arthur Carhart published a scathing indictment of the federal government's use of poisons, especially the 1080 it had developed to wipe coyotes off the continent. Carhart's "Poisons—the Creeping Killer," appeared in *Sports Afield* magazine three years before Carson's *Silent Spring* saw print. The piece even included passages about DDT, but Carhart focused most prominently on the poisons the government had developed to eradicate coyotes and their tendency to kill all along the food chain. "An area equal to one sixth of all the crop land in this nation is now being treated with deadly new poisons whose total effects are dangerously and shockingly unexplored," he wrote.

In a story in a magazine read by hunters and fishermen, a caption for a photo of a coyote killed by 1080 minced no words: "Agonizing

Taxpayer-funded Wildlife Services
still kills 80,000 coyotes a year.
Courtesy Dan Flores.

death throes of coyote killed by 1080, one of the deadliest poisons yet devised, show in snow furrowed away by thrashing of head, legs." In 1958, Carhart told his readers that for two decades a federal program had spent $500 million to spread the new poisons across 100 million acres in the United States, producing "a parade of death." Carhart was a tough westerner, but he ended his exposé with a powerful declaration: "I am scared."

Carson was scared too. She had become aware of and increasingly alarmed about DDT for the same reason that the mammalogists in the 1930s had called out the predacides in use then. Like the strychnine used against coyotes, DDT, commonly applied though aerial spraying, produced enormous collateral damage among nontarget species. DDT had first attracted attention for its success against typhus-carrying lice among World War II troops. After the war the chemical companies widely advertised it as a modern miracle, the solution for everything from agricultural pests to backyard mosquitos. They told a trusting public the spray was utterly harmless, and evidently we believed them. To a kid like me, growing up in sweltering Louisiana summers, only Jesus appearing in the clouds would have been more miraculous than the visits of the "spray truck." Suddenly it was possible to play outside in the summertime without becoming a mass of red welts from bayou mosquitos, and since everyone said it was safe, my friends and I rode our bicycles blissfully in the wet, white spray behind the trucks that rumbled regularly through town.

About the time my mother was making me change my DDT-drenched shirts before coming to dinner, Rachel Carson discovered that the chemical sprayed over a wildlife sanctuary near her home in Maryland was not just killing insects. Birds, in particular, were dying horribly in the days following spraying episodes. As she began to work on what became *Silent Spring*, Carson became convinced that by poisoning nature, we were not just fouling the environment around us; we were poisoning ourselves. As sensitive as the author was to the impact of poisons on nontarget creatures like birds, this was not Aldo Leopold's truly revolutionary biocentrism. Carson's major point was that human bodies are permeable and that DDT sprayed on insects bioaccumulated up the food chain to humans. Ultimately this anthropocentric argument made *Silent Spring* the revolutionary book it became. As she said, "If the Bill of Rights contains no guarantee that a citizen shall be secure against lethal poisons distributed either by private individuals or by public officials, it is surely only because our forefathers could conceive of no such problem."

Despite the chemical and agricultural industries' best efforts to portray *Silent Spring* and its author as "hysterical," within a year the book had gone through three printings, and some forty state legislatures were entertaining pesticide bills. Eventually it produced a ban on DDT use in the United States. But in a larger way, Carson's book aroused in the American public a growing revulsion at the ill-considered poisoning of the natural world taking place around us every day of the week. As she put it, misguided government policies were placing "poisonous and biologically potent chemicals . . . indiscriminately into the hands of persons largely or wholly ignorant of their potentials for harm." And much of that poisoning was in the service of the outright eradication of species after species, as if somehow North American evolution had floundered along for eons just waiting for us to show up to weed the place properly.

That kind of thing had been happening in coyote country for half a century already. But Leopold and Carson, and Walt Disney too, were making the world a different place in the 1960s. In 1963 John Kennedy's secretary of the interior, Stewart Udall, appointed a board to report on wildlife management in the national parks and asked it to look also at federal predator-control policies, particularly as they involved the use of poisons and especially on the public lands managed by federal agencies. One of the committee members, A. Starker Leopold, son of the now late famous author and biologist, had spent the last few years studying grizzlies in Mexico, where he and his graduate students had discovered that poisons put out by PARC Fish and Wildlife Service personnel threatened the bears' very existence.

The board's report to Udall, famous back in the age of ecology as the "Leopold Report," fell under Secretary Udall's eyes in the spring of 1964. Various constituencies over the years have either hailed or decried it. Among certain cognoscenti it is probably best known now for its historically naive assumption that Europeans had inherited a "wilderness" continent when they began colonizing America. Despite at least 15,000 years of Indian life in America and the obvious manipulation of continental ecology that implied, the Leopold Report recommended that the national parks be considered wilderness vignettes and preserved in the timeless condition they enjoyed when white eyes first fell on them. Postmodernist academics refer to this, in classes full of earnest graduate students, as "privileging a point of view."

Thirty years after the Leopold Report, I had several private breakfasts with Stewart Udall in Missoula, Montana, where I taught at the university, and asked him about the Leopold Report's seeming blindness to North American history. I recall that the former interior secretary paused for only a moment, then reminded me of how innocent the country had been in so many ways in the early 1960s, to the point that we barely considered that America even had a history before Europeans had arrived. "We seemed to think back then that somehow, some way, things like nature and evolution had gotten repealed by the twentieth century." He shook his head. "We thought we didn't have to pay any attention to nature anymore. The sixties pretty much disabused us of that!"

We haven't remembered the Leopold Report's predator recommendations in quite the same way as its account of the state of the national parks, but they were part of a string of events that were leading in a surprising and in fact revolutionary direction. The report's comments about the federal predator-control program were entirely damning, although in 1964 they fell short of calling on the government to end poisoning. Still, the board was willing to tell Udall that "the program of animal control . . . has become an end in itself and no longer is a balanced component of an overall scheme of wildlife husbandry and management." The report continued, "In the opinion of this Board, far more animals are being killed than would be required for effective protection of livestock . . . [and] wildlands resources."

While the Leopold Report's criticism of federal predator policy was important, it was not enough. As the age of ecology began to take off, it was becoming clear to scientists and the ever-growing legions of environmentalists that nothing would change unless change was forced. The agency's victims continued to pile up alarmingly. Within months of the Leopold Report, PARC's field agents had poisoned off, in Arkansas, one of the last remaining red wolf populations in the country. They treated the occasional Mexican wolves straggling across the border in the Southwest like scouts for terrorist revolutionaries and promptly poisoned them too. In 1965 one of the last populations of black-footed ferrets on the Great Plains disappeared when PARC operatives poisoned into utter oblivion the huge South Dakota prairie dog town they inhabited. Meanwhile, in California, agency use of Compound 1080 against coyotes produced a classic poison overreach, almost wiping the giant California condors off the planet. One condor carcass was so toxic with 1080, it killed all the beetles a museum used to try to strip away the flesh!

Starker Leopold's father, Aldo, had argued fifteen years earlier for a revolutionary principle in human affairs: a recognition that other species in this world possess an innate right to existence. "Biocentrism" in one sense was actually evolutionary. It implied yet another extension of the circle of ethical treatment that had begun long ago in human affairs, when we first moved beyond kinship and our own genetics and granted rights to others outside our families. In Western civilization, steps in that direction had included the Magna Carta of 1215, the constitutions produced by the American and French revolutions, the Emancipation Proclamation of 1863, the Nineteenth Amendment giving women the vote, and civil rights legislation in the 1960s. The step Leopold's followers pressed for in the late 1960s, however, struck many as the biggest step in extending ethics in human history. It advanced the radical idea that we offer ethical treatment—at least by guaranteeing their right to coexist on the planet with us—to other species.

The sea change underway in so many aspects of American culture was in full roar by the mid-1960s. Environmentalism would embrace a host of primarily human-centered issues: air and water pollution, toxic wastes, nuclear power, a search for renewable energy. But the ten years after 1964, following Congress's passage of the Wilderness Act, were truly the age of ecology, the most biocentric decade in American history. Leopold's and Carson's books and Farley Mowat's *Never Cry Wolf* initiated it, pop culture coyotes contributed in a kind of TV-land Rorschach suggestion, and an emerging sense of local uncoupling from the designs of the nation and a suspicion of authority carried it forward.

Accidental eradication of creatures it had taken North America millions of years to produce because we were too self-absorbed to notice was not a new thing under the sun. Maybe condors and black-footed ferrets and eagles were collateral damage in the same way that ivory-billed woodpeckers once were. Adapted to life in old-growth forests, ivory-bills ended up occupying too narrow a niche in the modern world. They disappeared not because of direct attack but because their habitat got logged. Of the eighteen mammal, thirty-four bird, and nine fish extinctions in America since 1600, some were "accidental"; species with small populations or very specialized niches, like ivory-bills, had simply died out. Most were victims of crass greed: market hunting destroyed the Labrador duck, the great auk, and, against all odds, the passenger pigeon, and the capitalist market wiped nearly 30 million bison and 15 million pronghorn antelope off the face of continent. But many other species fell into the same category as wolves and coyotes: they were coldly marked for outright extermination. We successfully eradicated the bright green and yellow Carolina parakeet, our only native parrot and one of America's most beautiful birds (look at Audubon's painting of them sometime) because, as with coyotes, agriculturalists thought they were pests whose lives weren't worth the space the creatures were taking up.

During this unique and special period, a wave of ecological mindedness was building to a crest. In 1964 Secretary Udall's office had compiled a list of sixty-three American bird and animal species that scientists believed were "rare" or "endangered," a number that grew to eighty-three by 1966. Udall called the bill drawn up by the

Lyndon B. Johnson administration to redress fears of extinction for these creatures the Endangered Species Preservation Act. Introduced into Congress by Representative John Dingell of Michigan, the act represented a couple of small first steps. It established the legal category of "endangered species," a list of which a group of international scientists was already compiling in a so-called Red Data Book. But the 1966 law made killing such species a crime only in a very circumscribed area: the US National Wildlife Refuges. Congress passed the act with little fanfare in 1966. The same blasé approach characterized 1969's Endangered Species Conservation Act, which also came out of the Johnson administration and made fish, crustaceans, and invertebrates—not just birds and mammals—eligible for "endangered" classification.

By 1969 Richard Nixon was president of the United States, but remarkably that did not mean environmentalism's moment in the sun had passed. Difficult as it might be to imagine from the perspective of the twenty-first century, environmentalism in the late 1960s was a bipartisan issue that at least some Republicans endorsed. When Nixon took office, the first Earth Day was only a year away. Anyway, saving the planet hardly seemed controversial in an age of inner-city riots and massive student protests against the war in Vietnam. Nixon himself, of course, had not the slightest interest in nature. On a spectrum of nature-loving American presidents, with Teddy Roosevelt and Thomas Jefferson occupying one end, Nixon pretty much bookends the other. But he recognized a political bellwether, and even if he privately thought environmentalist interest in animals was pathetic sentimentalism, Nixon believed that if he and his administration publicly endorsed environmental causes, he might be able to swing the student and youth vote toward the Republicans. If, that is—as one of the president's advisors put it—he could "identify the Republican Party with concern for environmental quality." How controversial could environmentalism be anyway?

So, counterintuitively, the Nixon Republicans actually created the Environmental Protection Agency (EPA) in 1970, although this did not fool many Americans. When the first Earth Day in history was celebrated in April of that year, *CBS Evening News* anchor Walter Cronkite reported that the crowds were "predominantly anti-Nixon." In his pursuit of youthful environmentalists, the president clearly needed another issue. And for someone with so little animal magnetism himself, he came up with a most unlikely one. With magazines from *Field & Stream* to *Sports Illustrated* to the *New Yorker* then running exposé articles on the government's poisoning campaigns against coyotes and eagles, in a Coyote trick as delicious as Elvis joining Nixon's antidrug war as a snitch, the president suddenly determined that embracing the well-being of coyotes could improve his political fortunes!

In the wake of the early 1970s media firestorm, the Sierra Club, Defenders of Wildlife, and the Humane Society joined to sue Nixon's Interior Department. Their antipoisoning lawsuit argued, among other things, that since the federal poisoning program was taking place without meeting the new legal requirement to conduct an environmental impact study, it clearly violated the administration's own National Environmental Policy Act. Seeking an immediate injunction to stop the predator-control program in its tracks, environmental groups eventually agreed to drop the lawsuit if the Nixon administration ended poisoning on public lands by 1972.

So Nixon appointed a committee to look once again at a federal agency that critics claimed was going way overboard in poisoning scores of thousands of coyotes, eagles, and even bears. While this committee included A. Starker Leopold, this time it was headed by Stanley Cain, a former undersecretary of Interior, and this time the outcome was different. Across less than a decade, there had been a seismic shift in the country's worldview. The Cain Committee's report, submitted to the White House in October 1971, did not equivocate: the administration should at once ban Animal Damage Control from using poisons to control coyotes and other predators. A practice that had been routine since the early century and viewed askance by only a handful of scientists suddenly began to seem, in the bright light of the new worldview, not just inappropriate but downright repulsive.

In the case of coyotes, the "silent majority" Nixon always invoked in his political speeches actually did include the president himself, or at least it appeared to if you take at face value the text of a policy speech he delivered to Congress. Nixon's February 1972 address probably reflected more the nuanced sentiments of its author, Republican environmentalist Russell Train, than any values deeply held by Nixon. Nixon almost certainly had never read Aldo Leopold's *A Sand County Almanac*, but this early 1972 environmental address nonetheless invoked Leopold's biocentric thinking in explaining the sharp new detour in the administration's policies.

America had reached a new stage of civilization, our president told us. "This is the environmental awakening. It marks a new sensitivity of the American spirit and a new maturity of American public life. It is working a revolution in values." He went on to explain what this meant. "Wild places and wild things constitute a treasure to be protected and cherished for all time. . . . [T]he wonder, beauty, and elemental force in which the least of them share suggest a higher right to exist—not granted to them by man, and not his to take away."

Nixon had not flinched from a line about other species' "right to exist," and he didn't flinch from the predator question either. "The old notion that 'the only good predator is a dead one' is no longer acceptable as we understand that even the animals and birds which sometimes prey on domesticated animals have their own value in maintaining the balance of nature." The president did acknowledge that the administration was, in effect, joining this concern—"The widespread use of highly toxic poisons to kill coyotes and other predatory animals and birds is a practice which has been a source of increasing concern to the American public"—but he was now prepared to make it his own: "I am today issuing an Executive Order barring the use of poisons for predator control on all public lands."

Banning poisons in the coyote war and acknowledging wild animals' "higher right to exist" were not platform planks that Nixon would campaign on for reelection in 1972. Indeed, when he discovered, despite his professed embrace of the Age of Ecology, that young people and environmentalists still supported Democrat George McGovern,

Nixon's environmentalism faded as magically as it had appeared. To his credit, however, he did follow through on the policy promises he made in early 1972. Two presidential executive orders followed on the heels of the address, one banning federal agents from using poisons against mammalian and aviary predators and scavengers, the other outlawing the use of poisons on US public lands.

It was enough of a break from history to stun anyone. Strychnine baits that had produced lightning-struck coyote corpses in America for 120 years—banned. Thallium sulfate that had resulted in blind, hairless coyotes shivering to freezing deaths—banned. Cyanide coyote-getters—banned. Compound 1080, amazingly, banned. There would be no more 1080-poisoned coyotes running frantically, uncontrollably, until they dropped. In 1972, those kinds of things at least *seemed* to be at an end.

Most remarkably of all, the administration drew up a new law, the Animal Damage Control Act of 1972, whose purpose was nothing less than a repeal of the 1931 law of that name that had first proposed the radical step of totally eradicating "the archpredator of our time" from the continent where it had evolved. The Nixon administration was proposing to get the federal government entirely out of predator control and turn that mission over to the states.

The president's attempt to repeal 1931's Animal Damage Control Act became a bridge too far for his natural allies in the agricultural and ranching communities. In conservative and rural Republican communities, the shit hit the fan. An outraged Farm Bureau led the charge, insisting that Congress acknowledge Fish and Wildlife statistics indicating that in 1970 in the sixteen western states, predator losses to livestock averaged 6 percent of stock numbers at a cost of $21 million to the industry. It encouraged stockmen's associations, like those for western wool growers, to bring coyote-killed lambs to public hearings on the bill. Western congressmen, as they long had, became their shock troops in the debate, one testifying that he was sick of hearing about "cruelty to the coyote" because the coyote was "a bad animal, a destructive animal" and the public ought properly to be more concerned about the coyote's cruelty to its victims. Nixon's new Animal Damage Control Act passed in the House in 1972 but never came up for a vote in the Senate. It never became law.

But special-interest fury against the new antipoison, antieradication Animal Damage Control law unexpectedly smoothed the way for the real crown jewel legislation of the Age of Ecology in America. What the US Supreme Court would soon call "the most comprehensive legislation for the preservation of endangered species ever enacted by any nation" was waiting in the wings. The Endangered Species Act of 1973, biocentrism's ideal translated into legal language, is arguably the farthest-reaching environmental law passed during the entire two decades, spanning 1960 to 1980, of a remarkable body of clean air, water, wilderness, energy, and toxic waste legislation sometimes known as America's "Environmental New Deal."

Despite the uproar surrounding Nixon's presidential proclamations halting half a century of callous coyote eradication on the country's public lands, the new act seemed to glide to passage almost unnoticed. The Endangered Species Act of 1973 built on its two predecessors but went much, much farther. It was in truth a watershed in global history, legislating other species' right to exist for the first time anywhere in the world. Yet, for all its potential, it somehow escaped the notice of its natural opponents, passing in the House by a vote of 390–12 and in the Senate by 92–0!

Less than a century earlier, the United States had stood unmoved as one charismatic wild species after another had disappeared or almost become extinct under our hand. Except for a few scientists and activists, no one had swallowed hard and looked away uncomfortably when the Animal Damage Control Act of 1931 had targeted the continent's coyote population for extinction. Now, between July and September 1973, Congress had passed a law that, via an almost backdoor approach, made the erasure of America's coyotes through deliberate or even inadvertent means illegal and impossible. Of course the Endangered Species Act did not end coyote "control." Individual coyotes by the millions were still fated to lose their lives in any number of shocking ways, a continuing source of outrage and activism by grassroots environmental groups today. But as a species, coyotes and hundreds of other threatened and endangered animals had earned the basic right to exist in America.

No one anticipated in 1973 just how much a great many people would hate this law. The Endangered Species Act derived its power from three of its sections, which in turn became the source of the many controversies that would envelop it. Section 4 gave the secretaries of Interior and Commerce the mandate to list species or subspecies as either threatened or endangered based solely on the *best science available*. In language that shocked conservatives when they got around to reading it, Section 4 outright prohibited the use of economic factors to determine listings. Section 7 compelled federal agencies to halt any development project that might imperil a listed species. And Section 9 prohibited the "taking" of any endangered species, which courts subsequently interpreted to mean not just shooting or harming an animal but degrading an endangered species' habitat, even on private land. One historian of the act describes Section 9 as "perhaps the most powerful regulatory provision in all of environmental law."

At the time of the act's passage, coyotes, by virtue of their evolutionary biology and intelligence, had never come close to becoming a threatened or endangered species. Despite all the technology, chemistry, and Dr. Evil pathological inventiveness we had thrown at them—and more was to come, for sure—coyotes were the extremely rare American mammal species still beyond our ability to push to the edge of extinction. And even if someone, somewhere, finally came up with a fatal canine disease or another surefire way to erase coyotes from nature, after 1973 the Endangered Species Act made execution of any such plan impossible. But if only theoretically so for coyotes, the Endangered Species Act became America at its best and most noble for wolves, bald eagles, grizzly bears, and black-footed ferrets—and, infamously to some, for snail darters, endangered arachnids, various warblers and flycatchers, prairie chickens, spotted owls, and scores of other species. For the animals themselves, though, for hundreds of species, it really was "morning in America."

Long before accounts of coyotes swarming into cities or hybridizing with wolves and dogs in the South and Northeast became the dominant narrative of the coyote story, all the fireworks had exploded, primarily in the West. Despite twenty-first-century coyote high jinks in downtown Chicago or Midtown Manhattan, and despite the passage of the Endangered Species Act and the presidential ban preventing government agencies from inventing ever-more lethal poisons to exterminate them, the war on coyotes in the West actually never slackened. Those waging the campaign had to seek out new killing methods, and because environmentalism itself produced a wide array of new activist organizations that came to the defense of predators like wolves and coyotes, mass killings of coyotes no longer went uncontested by the public. But the war went on and still does.

In 1972 it was a toss-up as to whether environmentalists, the livestock industry, or employees of the Division of Wildlife Services (PARC had gotten a new name in 1965) were most shocked when Nixon banned the use of poisons for government coyote control. Livestock interests, for their part, reacted to the ban on poisons as if it were a new tax levied on hardworking sheep to fund a food stamp program for homeless coyotes. Maybe the clap-to-the-forehead winners were actually Nixon's fellow Republicans, who began crawfishing as soon as the American Wool Growers' Association called for congressional hearings on the poison ban. New president Gerald Ford was willing to pardon Nixon in the aftermath of Watergate, but apparently pardoning coyotes in the aftermath of fifty years of the most outrageous persecution was more than Ford could stomach. In 1975 and 1976 he issued two executive orders that effectively restored the use of cyanide "Humane Coyote-Getter" tubes (by then called M-44s) on American public lands. For the wool growers and the Farm Bureau, that was a nice sanity-is-restored gift, but for reducing the density of the country's coyote population, the agriculture community really wanted its 1080 superpoison back. Thinking in more nuanced terms, the agriculture community also thought that the business of coyote killing properly belonged in the Department of Agriculture rather than in Interior. It turned out none of those was a wish too far.

Ford's "sanity" about coyotes helped win the interior West for him in 1976, but Democrat Jimmy Carter took the rest of the country and the presidency. For the coyotes that meant a reprieve and, in 1977, even a ban on federal field agents killing pups in their dens, an old practice that usually employed fishhook wire and forced smoke or gas—sometimes even dynamite—and seemed especially disturbing to an age of ecology public. But when the GOP regained control of the government with Ronald Reagan's election in 1980, the New Right launched a full-on backlash against what it called "the specter of environmentalism." Indulging a view that environmentalism threatened to replace communism as a primary threat to free-enterprise capitalism, the Reaganites were happy to let the puppies get their just desserts, the disturbed enviro public be damned.

So in January 1982, Reagan signed an executive order that not only overturned Carter's ban on killing coyote pups in their dens but reversed Nixon's landmark poison ban from ten years earlier. Three years later, in 1985, he further rewarded the livestock industry for its support by retaining the Fish and Wildlife Service in Interior but transferring its predator-killing operation to the Department of Agriculture's Animal and Plant Health Inspection Service (APHIS), a bureau for "protecting animal health and animal welfare." Reagan had granted the full wish list. Since 1976 APHIS's new predator agency had been called Animal Damage Control, which like Predatory Animal and Rodent Control had the virtue of unmistakable directness. But in 1997, in a move that has provided wince fodder for journalists and environmentalists ever since, Agriculture would return to the name Division of Wildlife Services for its predator-killing agency. Wildlife Services was a harmless-sounding label apparently designed to fool the public. Ah, this is who you called if you needed accent swans or a whitetail doe and fawn for an outdoor garden party. But nothing fundamental was different, and the name change to an innocuous I'm-from-the-government-and-I'm-here-to-help title did not hide from the cognoscenti that Wildlife Services was a frontier subsidy carryover whose primary mission was still killing.

Environmentalists got outvoted in the Ronald Reagan–George H. W. Bush years, but they were not so easily outmaneuvered. They could now call on the Endangered Species Act of 1973, since coyote poisons blanketing the countryside in the manner of the 1920s to the 1970s would inevitably threaten any number of species on the endangered list. By 1988 the courts had upheld that argument. Reagan's executive order may have raised shouts of exultation in the ranching community that 1080 was back, but Nixon-era environmental programs ensured that Reagan's own version of "morning in America" would not include a return to scorched-earth coyote poisoning. Ultimately, M-44 cyanide tubes remained a legal weapon for federal coyote hunters, but in 1985 the EPA approved 1080 only for use in plastic sheep collars that, when punctured, presumably poisoned just the coyote attacking the sheep wearing it.

Since the courts had now ruled that carpet-bomb poisoning the world with 1080 violated the Endangered Species Act, by the 1990s newly labeled Wildlife Services was at something of a loss for how to carry out its coyote-killing mission. But in truth, it had already invested decades in perfecting a new anticoyote technology anyway. With a little tinkering, an old southwestern story works to explain Wildlife Services' resolution of its dilemma: how to continue mass-killing coyotes for the ag community without mass poisoning. If the Indians had a coyote problem, according to the story, they'd send out one guy in a pickup. The Hispanos would shrug fatalistically: What can one do about a coyote? But the folks at the federal agency had a different reaction. When God (so the story always goes) looked down and said, "Look at those poor, ignorant Wildlife Services people. I gave them the rifle, the snowfields, and the airplane, and they didn't have sense enough to put them together," they looked at one another in wonder, imagined coyotes trying to escape pursuit from above, and bought some airplanes.

You do not just happen upon Wildlife Services' Predator Research Facility while out for a Sunday drive. No doubt by design, it is tucked away in the already semiremote Cache Valley in northern Utah, the final destination on a dead-end road a few miles south of the university town of Logan. No officious government signs point tourists toward it. Following the route in from the nearest highway required half a page of detailed instructions. Back in the 1990s, though, People for the Ethical Treatment of Animals still found it.

Julie Young, Predator Research's current director, and research biologist Eric Gese had agreed to a visit and sent the map, so on a fine, sunny afternoon in November, Sara and I drove over from the other side of the Wasatch Mountains, where Sara is a professor at one of Utah's universities. Since our Yellowstone trip she had become more and more intrigued by the coyote story, and there was no way she was going to miss this.

I grasped enough about the facility going in to know it was built in 1973 by coyote specialist Fred Knowlton (still famous for an adage in everyday use at Wildlife Services: "Coyotes will make a liar out of you every time"). That dates the Predator Research Facility to the moment when Nixon banned poisons for use in the coyote war. As the research arm of Wildlife Services, its mission harkens back to some pretty sinister roots—those original Eradication Methods and Control Methods labs of the 1920s are its spiritual antecedents—although it is currently a hub of the National Wildlife Research Center in Fort Collins, Colorado.

I knew, too, that the facility housed one hundred coyotes that were the subjects of research by biologists and grad students from around the world. Then there were these facts: Wildlife Services quietly spends $140 million of our taxpayer dollars a year not to serve wildlife but primarily to subsidize agribusiness. And its annual kill of

Wildlife Services' Predator Research Facility, successor to earlier federal extermination labs, secreted away in a mountain valley in Utah.
Courtesy Dan Flores.

all animals has ranged from a shocking 4 million in 1999, down to about 1.5 million a year in the early 2000s, then back up to a staggering 5 million in 2008 and well past 4 million in 2013. For their part, coyotes have never fallen out of Wildlife Service's crosshairs. From 2006 to 2011 the agency "retired" 512,710 of them.

I also knew that sheep raising, an industry federal predator control has forever existed to serve, has almost dwindled away in America. There were 56 million sheep on US pastures in the 1940s. There are fewer than 6 million now.

Coyotes are a political topic, and Julie and Eric are not quite sure what to think of our visit, but as good government servants, they are engaging and forthcoming. Fit and outdoorsy, with short dark hair and younger than I would have thought, Julie was born in California but graduated high school in Texas, went to Texas A&M, then on to a PhD as a carnivore specialist at Utah State. Slender, older, grayer— he looks to have spent much time in the field—Eric is an Arizonan who did his graduate work at the University of Wisconsin and worked with wolf guru Dave Mech. Both Julie and Eric have done extensive work on coyotes. They give us an entire afternoon, and I find I like them both. It also doesn't pass notice that in this facility the animal we are all interested in is a *KI-oht*. No doubt everyone Julie and Eric work with disdains a three-syllable animal.

This is an agency with a history that, to put it mildly, hasn't impressed the American environmental community. The *New York Times*, the *Sacramento Bee*, the *Washington Post*, and the journal of the Conservation Biology Society had all done recent, unflattering stories on it. There were petitions attempting to defund it, and half a dozen environmental groups had, a few months before our visit, joined a lawsuit against it. So naturally I have some questions as we gather around a table to talk, and I start out with what I think will be a softball lob to give them a chance to show their awareness of the criticism out there, that they're sympathetic.

"So, Julie, your tools for controlling coyotes now are obviously a lot more limited than they were forty years ago. Has that got your research going more in the direction of nonlethal control?"

Julie is inclined to a bit of a nervous laugh. "Well, not 100 percent. We're still working on lethal tools too. We've still got M-44s and of course leg-hold traps. And aerial gunning."

Aerial gunning, which really took off after Nixon's poison ban in 1972, dates to the 1940s as a federal antipredator tool. Wildlife Services is in love with it. Since 2001 its hunters have killed roughly 35,000 coyotes a year from planes and helicopters, images of which are never appealing when they make it to the media. I'm trying to find out whether the scientists at this Predator Research Facility are coming up with new, workable, nonlethal strategies for a new century. But, surprisingly, Julie deflects the question. She does want me to realize that regardless of the tools, the focus, at least, is narrowing.

"If we're looking at something lethal," she offers, "it's to decrease nontarget take, to get the problem animal rather than going after the whole population. Actually, a lot of what we do here is basic ecology in order to understand coyotes better, because you can't create tools if you don't understand the animal."

"Right, but back in the day the newest techniques for going after coyotes were always new poisons or, in the 1970s, aerial gunning, all developed by your predecessors at these research labs." I exchange a glance with Sara, aware that she knows where I'm going with this. On the drive over we'd talked about another agency, the Forest Service, once cut off from its more progressive constituencies by entrenched good old boys interested only in getting out the timber cut but then changed with the hiring of women and a younger, more environmentally savvy generation. I'm trying to lead us to the nonlethal topic again, hoping to hear that Wildlife Services is finally coming up with strategies that will get it off the hook some with the environmentalist public. "What cutting-edge techniques are you guys coming up with these days?"

Now Julie and Eric exchange a glance and some laughter. There's a bigger story here, and it's clearly Eric's domain.

"Well," he says, "what we're trying now goes back to the 1980s, when research showed that if you removed the pups from a coyote den, lamb losses went way down. It seemed to us as if provisioning the pups was a motivation for higher kill rates on lambs. When I got here I wanted to take that to the next level. We started with an experiment in the field where we had a vet sterilize half the coyote packs in our study. At the end of three years we found that our sterilized packs were only killing 12 percent of the lambs that packs raising pups were killing. We had some packs that stopped killing lambs altogether. So that's been our big idea." He lets that trail off, then laughs to himself about what he's about to tell us.

"The next step was, Can we implement this? Can we find ranchers who will work with us? But we just couldn't get anybody to bite. They felt like they already have the solution—we can aerial-gun, we can trap, we can use M-44s, so this is cute, but there's no need for it."

I'm mulling this information. I'm pretty certain the relentless pressure agency hunters put on them is one goad to coyotes taking lambs. They have larger litters to compensate and need more calories to rear them. But I ask, "So how expensive is it to sterilize? And how do you maintain a population of the coyotes themselves if you're sterilizing them?"

"We thought of that," Eric says. "But our idea is to make it surgical, just aim it at problem coyote packs. It's pretty expensive, $400 to $500 per coyote to catch them, do vasectomies on the males, and tie the tubes of the females. But over three years you'd come out even or ahead because of the lambs you'd save. We never even saw pair-bond changes among the coyotes; sometimes they still excavated dens together. Just getting the ranchers to try it has been the main problem. They're used to getting our help nearly free, so there's no way you could ask them to pay for it."

I was remembering a story I'd heard about a Wildlife Services public hearing in Idaho to explain sterilization. A rancher at the back of the room had raised his hand: "Son, I don't think you understand our problem. Them coyotes ain't fucking our sheep, they're eatin' 'em."

Julie brings me back to the present. "So we looked at cheaper, chemical methods. Last year we tested eighteen pairs using a drug that works extremely well on dogs. The male coyotes got ten times the dog dose, which, as long as they were alone, pretty much killed their sperm production. But as soon as they were reunited with their mates, fifteen of the eighteen pairs still got pregnant!" Julie still found this incredible. Sara and I do too. "That's just not supposed to happen, but somehow they're overcoming the drugs as a result of their social bonds. They even had normal-sized litters. Coyotes will make a liar out of you every time!"

"Anyway, right now, even if we had a viable chemical, the delivery system could be problematic," Eric says. "If it went out as bait, it would be hard to target the problem pack; plus there are endangered animals out there like swift foxes that could ingest a bait. We'll be testing this for a long time before we'll ever get EPA approval."

I nod that I understand that nonlethal controls apparently aren't on the docket for Wildlife Services anytime soon. "So in the meantime it's aerial gunning, M-44s, and trapping." They'd not said anything about 1080 sheep collars, the only way the infamous predacide can be used anymore, although—in another classic of coyote deduction—the more poison collars have gotten used, the more coyotes have figured them out. The hip new way for a coyote to attack a sheep is to hamstring it from behind rather than, as instinct would dictate, to grab the neck and crush the windpipe—and perchance puncture a 1080 collar.

"Yep, that's pretty much it." Then Eric adds, "With the toxins gone, the field guys are just trying to solve problems for our cooperating ranchers and counties. They're trying to be surgical."

"Problem solvers," Julie echoes. "We can't *require* our cooperators do nonlethal techniques anyway."

"Well, OK then, what's the annual *surgical* coyote kill for the agency these days?"

There's a bit of a pause.

Eric answered, "Well, I think the last couple of years, about 70,000 a year."

I actually already knew the ballpark figure, but 70,000 to 80,000 coyotes a year that are problematic enough to fly a goddamn plane off after them? "Given their replacement biology, what's the logic of that? This seems like it's never going to end."

"The way the ranchers see it, as long as you're taking some of them, you're reducing the risk," Julie insists. "And there's just the social thing of believing something is being done, a psychological thing."

"It has to be something they can see," Eric says. "Sterilization seems like hocus-pocus—you can't see anything—but shooting . . . and even nonlethal stuff like guard dogs, guard llamas, burros also work, and those are things ranchers can see, a visible effect."

I decide to direct one more question at Julie before we go out and drive the grounds of the facility. I've talked to enough coyote biologists by this point to know that studying coyote personalities is cutting-edge and that some of the boldness/shyness studies have been done here. "So, Wildlife Services seems to base much of its reason for existence on the proposition that there are always going to be problem coyotes because of the individual variation among them." The question comes out as more of a statement, but Julie confirms it.

"That is what I believe. They're individuals. Like people, some get into trouble."

We spend the next half hour navigating a pickup around and through the Predator Research grounds, sometimes stopping and watching the coyotes pace the perimeters of their pens. The facility has wire-mesh pens housing a population of one hundred adult coyotes divvied up into mated pairs, fed to keep them around twenty-five pounds, vaccinated, and treated for heartworms. They look healthy, if somewhat small, to me. On the couple of stops we make—the mountain valley setting of this facility is absolutely stunning on this late fall day—coyotes glide by, their golden eyes fixed on us; they move like water flowing in every direction and seemingly at every possible distance. And all of them are watching.

Sara asks what turns out to be our final question. "Do you guys have a relationship with any of the environmental groups, like Project Coyote for instance, or just grazing associations?"

I am in the front seat and can't see Eric, but I see Julie roll her eyes.

One of a hundred resident coyote inmates at
Wildlife Services Predator Research Facility.
Courtesy Dan Flores.

"Our greatest advocate is the American sheep industry. That's who we work with. As for Camilla Fox and Project Coyote, they and a bunch of others have filed a lawsuit against us, about forty pages long. I just deleted it. I'm not a lawyer, and I'm not going to waste my time."

Among the vocal defenders of coyotes across American history, the naturalists and the scientists are easy to pick out, people like Elliott Coues, Joseph Grinnell, and Olaus and Adolph Murie. There was the filmmaker Walt Disney. Then there were the writers: Ernest Thompson Seton, Enos Mills, J. Frank Dobie, and Edward Abbey, although the latter's shout-out to coyotes in *Desert Solitaire* about whether coyotes are eating enough sheep—"I mean, enough lambs to keep the coyotes sleek, healthy, and well fed"—was more a trademark Abbey thumb in the eye of western ranching than anything else. With the poison controversy of the 1960s and 1970s, though, modern nonprofit environmental organizations and their presidents or directors became the ones pushing hard to relieve the unrelenting pressure on predators like coyotes. The Fund for Animals was such a group. Rodger Schlickeisen, longtime president and CEO of Defenders of Wildlife and a man the western environmental newspaper *High Country News* once called "the most influential conservationist you've never heard of," was prominently another. Founded in 1947, Defenders of Wildlife has been among the most ardent advocates for coyotes since the 1960s.

So at times have more mainstream groups, like the Sierra Club, the National Wildlife Federation, and the Audubon Society. But more recently, niche environmental groups like the Animal Legal Defense Fund, the Center for Biological Diversity, and the Natural Resources Defense Council have been advocating for the rights of coyotes. In particular, a San Francisco–based organization called Project Coyote, founded in 2008 by Camilla Fox, has adopted a novel approach to taking on Wildlife Services, as well as demonstrating opposition to a highly controversial kind of coyote "event" that has become increasingly common across the country.

The latter would be the sponsored public coyote-hunting contest, with prizes (and often gambling pots) based on body counts for the most coyotes the contestants can, basically, murder over a week or a weekend. Coyote contests are often held by gun stores or local sportsmen's groups whose patrons are all in on the premise that failing to kill an elk or deer last fall has everything to do with coyotes (and in the West, now wolves) on the landscape. In 2000 the Fund for Animals financed a documentary on the practice, *Killing Coyote*, made by two filmmaker friends of mine from Missoula. It's not for the faint of heart. Coyote prize contests are advertised as family and youth events across the West, and despite national and international outrage, they are showing up in places like Pennsylvania. In Camilla Fox, though, coyote contests have a formidable opponent.

In 2013 I was in Northern California to do a talk on coyotes at the elegantly preserved nineteenth-century Nevada Theater in Nevada City. After a drive across the Central Valley, the visit gave me a chance to sit down with Camilla Fox at an outdoor restaurant in Larkspur, her home on the north end of San Francisco Bay, and find out why she had become such a passionate advocate in the modern incarnation of the coyote war.

Camilla Fox is striking. She is attractive, also charismatic, also tiny. She delivers information calmly and with the care of someone raised in an academic family, or so I think as I listen, and I turn out to be right. Her father, Dr. Michael W. Fox, long a leading researcher in the field of canid ecology, had a canid research station connected to his position at Washington University in St. Louis. "As a little girl I grew up with a wolf in the house and lived with her for fifteen years," she told me. "My father was never closer to any being than to that wolf. I was just inculcated into all things canid at an early age."

When her parents split, she'd lived with her mother in Maine and ended up attending Boston University, majoring in women's studies rather than biology. Then in 2006, after several years of work with environmental nonprofits, she went to graduate school at Prescott College in Arizona to write a very specific thesis about what she'd managed to make happen in Marin County, California, around Wildlife Services and coyotes.

"This was the project where you persuaded Marin to drop its relationship with Wildlife Services, right?"

"Exactly. I had learned back in 1996 about this agency and its use of 1080 in sheep collars. The way it works all over the West is that county ag commissioners actually contract for Wildlife Services to come in. The county puts up 30 to 70 percent of the cost of bringing in agency hunters, and the federal government pays the rest. In Marin the whole county was completely clueless about this arrangement. It functions with this aura of secrecy because once taxpayers realize what they're funding, they won't stand for it; they're outraged. This was at a time when a California ballot measure banned the use of M-44s and 1080 collars. Our board of supervisors in Marin was shocked to find out what they were signing off on every year.

"So we met with our ranching community. Marin is a place where wildlife is important, and we sought common ground with them. It led to developing an all-nonlethal coyote program for the county using shepherds and guard animals and fencing, and often bringing the sheep inside enclosures at night, and to Marin breaking off its old arrangement with Wildlife Services. In five years we were able to show that lamb losses went from 5.5 percent under lethal control to 2.2 percent under our system."

At this point, I'm recalling Wildlife Services scientists' assertion to me that paid snipers still had to go after problem coyotes in the Marin system. Camilla had also mentioned that she'd not yet had time to publish her thesis on the Marin plan versus Wildlife Services in a peer-reviewed journal, a piece of information the people at the Predator Research Facility had wanted to make sure I knew.

But I ask, "So have other counties around the country followed Marin's lead?"

"There's been a lot of interest, and some have asked for our data, but so far only Sonoma County has implemented something similar."

"Have you interacted much with the people at Wildlife Services? It struck me visiting with their scientists that, unlike the Forest Service with its multiple constituencies, this is an agency with only one interest group, the agricultural community, which blinkers them to bigger realities. Their scientists may be myopic. They don't seem irrational though."

"Wildlife Services does all this great research at their centers," Camilla agreed, "but there's this disconnect in translating it into the field, to the guys who are most used to traps and airplanes. There's some grudging acknowledgment of that. I think there are a few people there who are younger, smart, well intentioned, but they get burned out from the inertia of the old killing system. There's concern, too, that if Wildlife Services shifts to something different, somehow their funding's going to dry up. I think they have a great opportunity to shift their whole paradigm. But it would be a big cultural shift."

It was a gorgeous autumn afternoon on the bay. Bright sun, drifting cotton-ball clouds, light traffic tending toward BMWs and sports cars idling a few feet away. I had just one more question for her.

"Why coyotes, Camilla? Why spend your life fighting to protect them? The gods know they don't need much help from us, do you think?"

I'd spent twenty years among nonprofit environmentalists in the university town of Missoula, Montana, and I knew this was going to be a fat fastball down the middle for her, right at her group's mission.

"In a way, I'm doing this for us, for the sake of our moral compass. We've made coyotes the most persecuted native carnivore in North America. My board—lots of illustrious scientists and ethicists—estimates that considering all the ways we do it, we're killing half a million coyotes a year, which averages one every minute! But here is our iconic song-dog, so prominent in our history, so unique compared to wolves or foxes that occur on other continents. If we can change the way we view and treat coyotes, we can change the way we treat nature itself."

Within a few months of my meeting with her, Project Coyote had managed, among other things, to get public coyote-hunting contests

Poster for prize-based public coyote hunt, 2013.
Courtesy Dan Flores.

for prizes and gambling pots banned in California. This was going to be harder to do elsewhere, I knew, where "sport hunting for coyote control" strikes many as sanctioned by God. The real goal is to have convenient live targets, of course, but the argument often advanced is that gunshot coyotes make for more "game."

Of course, sport hunters don't read peer-reviewed ecological articles. Since the time of the Murie brothers' research, biologists have done some very fine-grained studies of coyote predation on the animals hunters like to shoot. The strongest case for coyote predation has long been the pronghorn antelope, as in some instances coyotes account for half the mortality of pronghorn fawns. Coyote predation on pronghorns, though, is complicated by livestock grazing, which impacts forage enough that antelope fawns end up weak and easy prey. Even so, antelope populations remain stable with that level of predation, in good part because female pronghorns long ago adapted to coyote pressure by birthing twins. This is a more ancient equation, for the coyote and the pronghorn alike, than we ever think to acknowledge.

True enough, in the East and South coyotes are now the primary predators of whitetail fawns, although their effect is actually to create healthier deer herds more in balance with the setting than before. In the West, where mule deer are an important game animal, an intensive, long-term study of coyotes and mule deer in 2011 concluded flatly that "benefits of predator removal . . . will not appreciably change long-term dynamics of mule deer populations in the intermountain west." In obstinate defiance of that science, in 2012 Utah passed a controversial Mule Deer Protection Act creating a predator bounty program that pays state hunters a $50 head-price for coyotes. In 2014, 7,041 coyotes paid with their lives for being in a state where neither hunters nor legislators bother to read science.

Meanwhile, in states like Montana, Wyoming, Idaho, and New Mexico, hunter-sponsored public coyote contests continue to go on, ostensibly "to protect game," often for prizes. Sometimes the body counts reach as many as two hundred animals over a weekend.

Along with other groups, Project Coyote continues to battle hunts like these, now with a documentary film narrated by Peter Coyote that may finally blow open the disgrace of these coyote contests, which will struggle to survive in the glare of national and international attention. There's no vantage from which they're not despicable. Project Coyote also brings lawsuits. In November 2014, a coalition it assembled, which included the Animal Legal Defense Fund, the Natural Resources Defense Council, the Center for Biological Diversity, and the Animal Welfare Institute, sued Mendocino County and Wildlife Services on the grounds that Mendocino had failed to

Coyotes discarded in the desert, the aftermath
of a coyote-hunting contest in New Mexico, in 2015.
Courtesy Kevin Bixby.

submit its $142,356-a-year contract with Wildlife Services to environmental review as required by the California Environmental Quality Act. In February 2015 Project Coyote filed a second suit against the US Department of Agriculture, this time seconded by four additional environmental groups, over the department's failure to conduct an environmental-impact assessment, as required by federal law, of Wildlife Services' extensive predator-killing program in Idaho, financed entirely by taxpayer dollars.

In 2015 Camilla Fox was named the John Muir Conservation Awards' Conservationist of the Year.

———

My first-grade class picture shows me decked out in Davy Crockett gear straight out of the 1955 Disney series starring Fess Parker as the great American frontiersman. By the time of the Endangered Species

Act of 1973, I looked more like one of the Indians the Crocketts of the world had displaced. Big History had unreeled in the years since I had matriculated from first grade, and pop culture was right there in the mix.

When it came to changing opinions about coyotes, Walt Disney was sure in the mix. A Chicago native, Disney had spent his early professional career in Kansas City, but in 1924 he and his brother Roy had transferred their operations and fortunes to Hollywood, where they quickly became commercial successes with cartoons and shorts in the motion picture industry. Disney may have fought to keep the Cartoonists Guild out of Disney Studios, and he may have been a red-baiting Republican in 1950s Hollywood and a Goldwater Republican in the 1960s, but he had always been interested in animals. His studio effectively invented the nature documentary, and he inserted conservationist values into many of his films. The house he built at Smoke Tree Ranch in Palm Springs in 1957 gave him more chances to be in the country and to interact with desert coyotes, and he soon began to hear from friends about the coyotes hanging out in the Hollywood Hills area of Los Angeles. These coyote experiences brought Disney around to defending the little song-dogs of the western deserts and cities.

The Coyote's Lament of 1961 was just the beginning of Disney's attempt to shape American attitudes about coyotes. He followed it later that fall with a theatrical release, *Chico, the Misunderstood Coyote*. For some moviegoers the star coyote's name probably sounded vaguely familiar, and for a reason. "Chico" was none other than "Tito," the heroine coyote in Ernest Thompson Seton's story from sixty years before. Perhaps the Bambi-like storyline appealed to Disney. *Chico* is the story of an orphan pup whose family has been killed by ranchers. Captured, he is put in a rundown roadside zoo, a classic highway snake pit in the West, where his captors display him to gawking tourists as a "wild desert dog." Disney's film even invokes the poisoning debate by having Chico nearly killed by poison. But as in the Seton story, Chico escapes with a newfound craftiness about what the human world holds for his kind.

Now Disney was rolling, and Chico was about to have new adventures. In 1965's *A Country Coyote Goes Hollywood*, a theatrical featurette narrated by and starring country music star Rex Allen, Chico gets chased inside a moving van by dune buggy toughs and their dogs and finds himself in Hollywood, learning how to be an urban coyote. The surprise plot twist is actually an indication of how hip Disney was to coyotes in modern America, for Chico finds that Los Angeles is already inhabited by a population of slick, big-city coyotes. The film includes outstanding footage of coyotes in the Hollywood Hills. Disney even released a 45-rpm record from the film, "When Coyotes Howl in Hollywood You Hear a Mournful Tune."

Throughout all those pivotal Age of Ecology events in the late 1960s and early 1970s, *Disney's Wonderful World of Color* kept at it. *Concho, the Coyote Who Wasn't*, about a young Navajo shepherd who trains an orphan coyote to work as a sheepdog, appeared in 1966. *The Nashville Coyote*, about a coyote with a voice for the stage, came out in 1972; and *Carlo, the Sierra Coyote*, based on the novel *Sierra Outpost* and with John Muir as a character, debuted in 1974.

Unlikely as it might have seemed, then, Uncle Walt deserves some share of the credit for helping the country begin to cast aside the disrespect for coyotes that started with *Roughing It*, then morphed into outright hatred via the public relations arms of the Bureau of Biological Survey and its descendants. The gray wolf's emergence as environmental star helped give coyotes a fresh chance too, but that never could have happened without the American public's own maturation beginning in the 1960s.

And it seemed, somehow, as if the coyotes mysteriously understood that maturation, for at this same moment they discovered a new refuge for themselves in America, one chock-full of food and cover where, blessedly, no one *ever* shot at you.

Hello bright lights, big cities.

CHAPTER 6

Bright Lights, Big Cities

It is a brilliantly sunny June day in 2013 in Eldorado, New Mexico, a subdivision southeast of Santa Fe that has long been a second-home and retirement destination for East and West Coast academics and movers and shakers from around the United States. About forty-five well-heeled residents have shown up for a 10 a.m. presentation on urban coyotes by Project Coyote, Camilla Fox's San Francisco–based organization, which has a local chapter in northern New Mexico. The centerpiece of the meeting is a documentary film, *Still Wild at Heart*, about how coyotes, after an absence of decades, in 2001 began to recolonize San Francisco. Urban coyote expert Stan Gehrt appears in the film to discuss how coyotes similarly moved into Chicago and tells the camera that he began his studies thinking there were maybe a few dozen coyotes in that city, only to conclude after a few years that the population was more than 2,000, with almost everyone in Chicago going to bed at night with a coyote no more than a mile away. (A year later, when I asked Stan about that figure, he equivocated on a precise number, but said, "It's a lot more than that now. A lot more. Chicago has become a source population for the surrounding hinterland.")

This is liberal northern New Mexico, so everyone, with the exception of Justin, the local boy who is the state coyote-mitigation officer, pronounces our canid's common name *ki-YOH-tee*. The theme of the gathering is coexistence. For those encountering coyotes in the backyard or along the trails in the local park, that means following some simple rules, a message that has acquired some urgency as more cities find themselves dealing with coyotes and as town coyotes have steadily become more comfortable around us.

The prime directive is straightforward and delivered with an exclamation mark: For chrissake, do not feed coyotes and accustom them to associating food with humans! To avoid the most common human conflict with coyotes, don't let your cats or small dogs outside at night. Don't leave infants or small children unwatched outside. And "haze" urban coyotes, who after a few generations of city life lose whatever fear of us they ever had, which probably wasn't much to start with. Whatever you do, don't let a streetwise coyote bluff you. If your dog or your jogging or biking excites an unusual and bold reaction from a coyote, establish your dominance. If a coyote doesn't retreat from you or acts in any way aggressive, stand tall, raise your hands over your head to underscore the fact that you're a hell of a lot bigger than it is, and shout. If you've got a good arm, pick up a couple of rocks and prepare to deliver a Nolan Ryan dust-off fastball. Give the coyote every indication that you're fully prepared to kick its little ass halfway to Sunday. In other words, keep town-wise coyotes thinking that people can still be dangerous. Or at least that we're way too weird to trust.

Those enlightened sentiments in 2013 were not necessarily where we started out when coyotes first came to town. Not by a long shot.

For a good reason, US towns and cities were hard for coyotes to enter until fairly recently. But their instincts about associating with humans were visible in other contexts. Nineteenth-century traveler accounts brim with references to coyotes as a constant presence around the camps of emigrants, trappers, traders, and explorers, where "prairie wolves" often dashed through camp to snatch something and run with it. The journals of Meriwether Lewis and William Clark are full of accounts of both coyotes and wolves almost underfoot. Clark once bayoneted a wolf strolling past him, not because it presented a danger but apparently just because he could. A writer for Salt Lake City's *Weekly Tribune* in 1887 felt the need to warn his readers about the coziness of coyotes on the trail: "Have you ever seen a coyote? He is the most impudent animal that exists. . . . [A]s a kleptomaniac he is an expert, for he can steal the boots from under a camper's head and the meat out of his camp kettle."

Simple canine competition alone kept coyotes from establishing territories and scavenging and hunting among us in places like Los Angeles, Albuquerque, Denver, and Calgary a century ago. Except in cities, dogs, not wolves, were the obstacle. The past can be a foreign country, and this may be one of those times, for few modern city dwellers can conjure in their minds an image of cities overrun with loose and feral dogs. But that was the case for most American cities and towns into the beginning of the twentieth century.

If we still lived in the loosely regulated cities that characterized American urban life 125 years ago, urban coyotes might be a rare phenomenon. Dogs and packs of dogs, a high percentage of them unowned or only casually linked to owners, once roamed at will through the streets of cities and towns, and as canine competitors of similar size, they kept coyotes at bay. According to historian Jon Hall, when word began to spread that ten Americans in the first few months of 1848 had died horribly from rabies contracted through dog bites, the so-called Great Dog War marked the beginning of the end for that older urban world. Led by New York, Philadelphia, and Boston, cities across the country proceeded to launch a violent, clumsy offensive against stray and feral dogs. Bounties, urban sharpshooters, dog clubbers, and even lidded cisterns designed to drown hundreds of captured dogs at a time became the order of the day, always to a certain middle-class distaste leavened by resignation at the inevitable. Eventually dogcatchers and pounds became the ultimate answers to the Great Dog War. The urban middle class may have ended up feeling safer, but the excesses of the cure were sufficiently egregious to help produce the Society for the Prevention of Cruelty to Animals.

In the end, the Great Dog War didn't just turn the remaining dogs into pampered, yard- and house-kept animal companions, ushering in a new and modern relationship between owners and their dogs. In one of those entirely unexpected ecological consequences, a dog regulation in the urban landscape opened up American cities to a new town canine of a far wilder and more exciting sort. America's junior wolf was about to become a denizen of big cities.

———

New York City may represent the newest frontier of Coyote America, but coyotes probably first became city slickers 3,000 miles in a desert direction, in the Los Angeles of the immediate post–World War II era. The City of Angels had, after all, been wrested from wild coyotes and their fellow travelers back when the Spanish padres had founded it in the late eighteenth century. There is some evidence that town building along the Los Angeles River never did entirely exclude wild coyotes from the city.

In the heart of southwestern coyote country, the same was probably true of San Diego, Tucson, Phoenix, Albuquerque, and Las Vegas. The progression was likely similar in all of them, beginning with the founding of a city in prime coyote habitat, which likely drove the wild canids to the margins of town. Several decades or a century of few or no dog regulations would have prevented coyotes from fully colonizing these cities, but then passage of laws against free-roaming dogs and the advent of dog pounds and dogcatchers unintentionally allowed coyotes to establish territories inside the city limits. Train rights-of-way that preserved corridors of habitat augmented colonization, allowing coyotes access to inner cores. The city parks movement preserved a few natural areas where coyotes could escape the human din of urban life, find prey, and raise litters. From the perspective of modern city dwellers, seeing small wolves trotting around town was at first lights-out shocking. From the point of view of the coyotes, adapted to the presence of human encampments for 15,000 years and human cities for 1,000 years or more, it must have all seemed entirely normal, effortless, and natural.

For contemporary biologists studying the urban coyote phenomenon, like Stan Gehrt, Seth Riley, and Stewart Breck, what began as far back as a century ago in the cities of the desert Southwest has now assumed a shape and direction, and the patterns are repeating in Seattle, Denver, Chicago, and now New York. Los Angeles—where a population of some five hundred town coyotes had become famous enough that as early as 1965 Walt Disney would make *A Country Coyote Goes Hollywood* about them—saw modern Coyote America first. But with only a few variations, coyotes and the human city dwellers among whom they now live are replicating those patterns from coast to coast.

One of the most common twentieth-century conceits in the Western world was that we humans, more urban by the decade and as a consequence seemingly divorced from the wild, were somehow "outside nature," separate, by virtue of our specialness, from the natural world. Put aside the folly of such an idea for a species arising out of Earth's evolutionary stream—whose very bodies are microbiomes of thousands of other species and who will never be separate from nature unless they figure out how not to die—and focus on the urban component of that idea. No human settings have struck us as more polar opposite to nature than cities. Cities were human conceived, we said, human designed and built. If modern city dwellers take anything for granted, it's that cities are surely the one place they don't have to engage with wild predators. Yet cities are ecosystems. They sustain what biologists call synanthropic species, or creatures that thrive in the ecosystems created by urban development. We are one of those obviously. Rodents are another. And so are coyotes.

To a coyote slipping along a rail line to enter a city for the first time, urban ecosystems probably only exaggerate the experience of living with humans in the rural countryside. Places useful to a coyote would be more scattered and broken up because of the prevalence of asphalt, concrete, and structures. As a coyote moved from suburbs to edge cities to inner cores, the din of noise we humans make would gradually crescendo. The world would become more lit, and the effect of lighting would extend through the nighttime. A coyote confronting big-city life would find a massive increase in the number of roads it had to deal with, and the number of cars on those roads would go up by several orders of magnitude. This would be especially true during the daytime, since as a species we are most active then.

A newly urban coyote would find mice of various varieties, a coyote's most dependable prey, to be extremely numerous, but there would be strange new prey, too, like Norway rats, urban-adapted flocks of geese and ducks, and exotic plants and fruits of wide diversity. While dogs would mostly be contained in modern cities, the occasional small dog let out on its own might arouse a coyote's ire as a potential predatory competitor, or a coyote with pups to feed might find it tempting. Cats don't suffer the restrictions that dogs do in most cities, and they would be numerous and sometimes look like prey to a coyote. They would also evidently strike an urban coyote as what biologists refer to as an "intraguild predator" that it should eliminate from its territory.

Modern city coyotes must have first figured out how to insinuate themselves into this much human density in Los Angeles, but the strategies for accomplishing it and the trajectory of how coyotes are adapting to life even in developed downtowns are now unreeling in towns and cities all over North America. New York City is just the most recent.

Stan Gehrt, America's go-to biologist for urban coyotes, who has been studying them in Chicago for fifteen years, characterized the urban coyote story to me this way: "It's an ongoing, unplanned experiment." As of 2015, understanding coyote individuality is a key part of his work on urban coyotes: "Right now we are really interested in coyote personalities more than we've ever been," he told me. Like people, some coyotes can take a lot more humanity and novelty than others. Coyotes calm enough to tolerate noise, traffic, lights, the

Amazon-like torrent of human scents, and the frequent sight of humans are the most successful in cities, although they can develop that behavior through "habituation." With only a couple centuries of evolutionary pressure in their past pushing them to avoid humans, urban coyotes living in the present don't find it so tough to lose their fear of us, it turns out. We think it normal for animals to flee from us in a wild panic, so the "habituated" coyote also gets into trouble in town. At least that's the case now. The goal, though, is for coyotes and modern city dwellers to learn how to "cohabitate"—in effect to create the sort of relationship that prevailed among coyotes and people, among coyotes and urban Aztecs—across North American history.

Coyotes with a tolerance for being around us are finding new riddles to solve in urban life, and coyote intelligence seems fully up to the challenge. Indeed, city life may well be selecting for novelty-seeking, "supergenius" coyotes. Among the most dangerous aspects of urban life for canid predators are the teeming highways they must learn to navigate. Coyotes can do this by moving mostly at night, when people are less evident and traffic ebbs; a nighttime routine has become one of their adaptations to city life. But Ohio State's Gehrt has seen them out in Chicago during rush hour, crossing half a multilane interstate highway brimming with traffic, then sitting in the median until it thins enough to cross the other half.

Coyotes are still getting hit by cars, but in another few generations we may discover that city life has fashioned urban coyotes and an urban coyote culture that deals with highway traffic with the skill of pedestrians in modern Rome. Los Angeles coyotes already seem to demonstrate this trend. In Chicago more than 60 percent of coyotes die under the wheels of cars. But with several more generations of city experience in carmageddon California, LA coyotes have gotten that figure down to about 40 percent. Angeleno coyote culture apparently has even recognized one highway as a barrier to travel: only the most intrepid California coyotes ever attempt to cross US Highway 101, which runs north-south through the state.

Tasked with establishing territories in the fragmented patchwork of parks, green spaces, campuses, golf courses, rail lines, and deserted

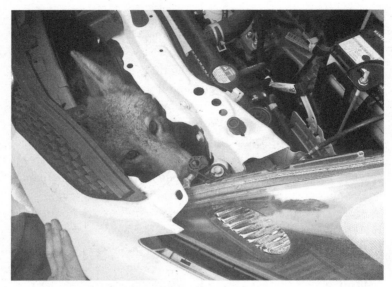

Coyote that survived high-speed car hit,
then hitchhiked to California in the grill in 2012.
Courtesy David Lovere/Rex Shutterstock.

lots that make up citified natural habitats, urban coyote packs generally create smaller home ranges than they would in rural areas. In Los Angeles, fifty-three radio-collared resident coyotes had an average home territory of about five square kilometers. The figure for 118 resident coyotes in Chicago was the same: five kilometers, or about three square miles. Coyotes in rural areas require larger territories, averaging about seventeen kilometers. (Solitary, transient coyotes in both urban and rural locales use far larger ranges of twenty-five to fifty kilometers for city coyotes and up to one hundred kilometers—sixty miles!—for rural nomads.) Biologists have concluded from this that urban coyote territories are far more resource-rich compared to those in the rural countryside. And smaller home ranges indicate a denser coyote population, meaning many more coyotes can live in a city of a given footprint than would be the case for a similar-sized territory in the country.

If it seems counterintuitive that a predator like a coyote would find life in town to be fat-city, consider this additional evidence: in rural Illinois, where residents shoot, trap, and harass coyotes, only 13 percent of coyote pups survive to maturity. In the Chicago metropolitan area, a whopping 61 percent of coyote pups survive to adulthood. Like human adolescents, male coyote pups are always the most at-risk pack members, the easiest to trap or poison or shoot. But in town young male coyotes tend to survive at the same rate as females. In fact, only in preserved wildlands like national parks does coyote survivability compare to what coyotes experience in cities. For a twenty-first-century coyote, town life is pretty obviously the good life, especially compared to the dangers of rural America. We're going to have to start imagining cities as twenty-first-century coyote preserves in much the way national parks were in the twentieth century.

Any version of a good life means eating well, one reason for being in a city in the first place. The emerging urban legend, propagated by a media clearly out of its depth when it comes to city predators, is that town coyotes eat very well indeed because they're dining regularly on small pets, pet food, and trash. That pearl of wisdom may be an urban memory preserved from the Great Dog War of the nineteenth century, when loose and feral city dogs survived by doing pretty much just that. The biologists have shown repeatedly that this is not how town coyotes survive. Individual coyotes or specific packs may develop a taste for human foods or pets, as has happened a few times when coyotes in Tucson, San Diego, and on the outskirts of Seattle developed a yen for cats (which in Seattle actually made up 13 percent of a particular pack's diet). The coyote-cat story, though, is much more nuanced than the proliferation of cat-attack accounts suggests.

Despite the cat-killer urban legend, in city after city the science indicates that pets provide only about 1 to 2 percent of the average coyote's diet. Stan Gehrt grew up as a rural Kansan who'd never seen a city the likes of Chicago in his life, but coyotes had attracted so much notice there by the late 1990s that he finally got funding to study them. Unless new isotope analysis of coyote diets shows something his original scat studies didn't, he told me, then "despite the stories, I can say flat out that urban coyotes don't depend on pets for food. If coyotes were relying on pets as a source of food, we quickly wouldn't have any pets left."

That may not translate, as Gehrt was quick to add, to a peaceable kingdom between urban coyotes and domestic cats and dogs. Coyotes may kill more cats than they ever eat. That may have to do with an ancient animosity; recall that naturalist Thomas Say finally collected the type specimen of the American coyote by baiting his traps with a bobcat. Coyotes may also attack cats for the same reason they attack small dogs: they perceive domestic cats and dogs as intraguild predators operating in their territories. When coyotes attack dogs or cats, they most often don't intend to eat them; they're simply ridding their territories of roaming predators. The vast majority of coyote encounters that unnerve urbanites almost always feature a dog. It's not true in every case, but in a large majority of instances of coyotes biting people, the bite happens while a dog owner is attempting to protect a pet.

What urban coyotes eat depends a good deal on the city where they live. In Chicago, the large lakeshore population of Canadian geese has become a major food source for Cook County coyotes, not so much the adult geese themselves as the contents of their nests, nearly half of which get raided in most years. Where there are populations of deer in American cities, coyotes can quickly become major predators of fawns. Coyotes acting as a control for urban populations of deer and geese, it turns out, is one of those "beneficial" outcomes Olaus Murie wrote about in the 1930s. Although not many cat owners will want to hear it, increasing numbers of studies indicate that when coyotes come to town and pilfer the odd cat, the survivability of local songbirds goes up markedly.

Compliments of the Los Angeles of thirty years ago, coyote dumpster diving is an urban legend with legs. Some biologists believe that as much as 25 percent of the diet of some coyote packs in LA in the naive 1980s was human food. More recent studies of urban coyote scat indicate that in most cities the percentage of trash, pet food, and other human food actually comes in at only about 2 percent. A recent study in modern Denver pegged that figure at less than one-half of 1 percent, and today it has dropped to 6 percent even in LA. Despite all the anecdotes from the 1980s, except in rare cases of localized coyote culture, the vast majority of town coyotes are not scavenging behind Sonic and Burger King. They're not really much of a threat to the six-pack of tallboys you left on the porch.

Sometimes, especially in summers when a coyote pair is stressed trying to raise pups, the parents might become serial killers of cats (one British Columbia coyote den yielded fifty-five cat collars). If you are halfway intelligent with your animals, though, coyotes are not remotely as great a threat to your cat or dog as traffic is. Coexisting with coyotes just requires paying attention, the way we've done around predators for a couple hundred thousand years, after all.

Still, coyotes are a kind of wolf. Living in our midst, are they a danger to us?

Los Angeles is famous among ecologists for having the most extensive wildlands-urban interface of any city in America, a zone at least seven hundred miles long where subdivisions abut chaparral and sharply incised canyons cut deeply back into mountain ranges like the San Gabriels, the San Bernardinos, and the Santa Monicas. Those canyons provide thousands of patches of natural habitat that interpenetrate the edges of greater Los Angeles. From the Coyote Hills to the Hollywood Bowl, from urban parks to university campuses, coyotes are everywhere in the six counties of greater LA. They probably always have been, but in the 1980s they began to attract attention for some of the same reasons Central Park's "Otis" would freak out New Yorkers in 1999.

As Mike Davis wrote in *The Ecology of Fear*, published a year before Otis showed up in Manhattan, in 1980s LA coyotes became "symbols of urban disorder," of a breakdown in our own conceits about what city life meant. We've thought of cities for 5,500 years of recorded history as the one spot where, at long last, humans could escape predators. But in modern America, it turned out, not so fast.

The Los Angeles of the 1980s became the place that wildlife managers in Denver, Seattle, Chicago, St. Louis, Cleveland, and New York are today doing their best not to emulate. Coyotes had been in LA for decades, only attracting passing notice; as far back as 1938 the city government paid bounties on 650 coyotes the first year bounties were offered. A population of at least five hundred coyotes was residing in LA by the 1960s. But the lure of the Southern Californian good life and the success of the sprawling late-twentieth-century city gave LA a population of people from across America and the globe, many of them recent arrivals who knew little or nothing about coyotes other than that one endlessly fell off cliffs in Saturday morning cartoons. In the 1980s, scientific studies of urban coyotes were still in the future. The majority of the LA population did know one thing about coyotes though: As wild predators they sure as hell weren't supposed to be inside the city limits of a giant metropolis. Yet there they were, trotting through the tombstones of local cemeteries, loping across the runways of Los Angeles International Airport, and, most disconcertingly, hunting along suburban streets where people lived.

The unease about coyotes in LA spilled over into panic on August 26, 1981. That morning, in a new suburb of Glendale, three-year-old Kelly Keen wandered, unattended, out of her house and into the driveway. A single coyote attacked, killing her. It was the first human death attributed to a coyote in recorded American history. Glendale officials responded by killing every coyote they could find and astonished Los Angeles when their efforts produced fifty-three dead coyotes in the square mile around the Keen home.

Los Angelenos' immediate, emotional response was to describe areas like Glendale as "teeming coyote ghettos" and to compare coyote packs to "gang bangers." Any coyote spotted in the daytime became a "brazen criminal," bold enough to show itself "in broad daylight." To writer Mike Davis, assessing LA's reaction fifteen years later, coyotes were "the textbook example of a protean, 'unfinished' species" that engaged in "continuous behavioral improvisation." Coyotes survived city life, he wrote, by eating garbage, pets, even zoo animals. The title of his chapter on the story of coyotes and cougars in LA: "Maneaters of the Sierra Madre."

Today biologists believe that more than 5,000 coyotes inhabit greater Los Angeles. Their territories so blanket the city, and they do so well there—living significantly longer on average than rural coyotes—that biologists suspect LA, like Chicago, has most likely reached its carrying capacity for coyotes and is actually producing a surplus number of animals that leave the city to find territories outside it. Plenty of LA residents still hate them, but in a pattern that urban coyote researchers are finding increasingly common, residents have slowly recovered from the initial shock of realizing they share their city with a small, wolflike predator. Over time, urban people get used to coyotes. They go Aztec and learn how to live with them, which essentially entails keeping coyotes wild and a little nervous even in the city. By now plenty of people with urban coyote experiences under their belt have come to relish the presence of coyotes in the city for ecological reasons or just because they're so beautiful and it's such a cool thing to get to see a small wolf among us as we go about our daily routines.

Wildlife managers respond to the winds of politics, and one bit of evidence that attitudes toward coyotes are beginning to evolve is that in cities like Los Angeles, Chicago, and Denver, officials charged with managing the relationship between urban coyotes and city residents have developed progressive plans of coexistence to replace the early kneejerk attempts to eradicate every coyote anyone sighted. Managers certainly still take out problem coyotes, the ones that are aggressive toward people or become dependent on pets as a food source, but in most cities these are still a tiny percentage of the total coyote population. No city wants to replicate the Los Angeles experience, where between 1960 and 2006 nearly seventy people were bitten by coyotes, accounting for fully half of all the coyote-bite incidents in

North America. Too many Angelenos, often unintentionally or due to ignorance, had fed coyotes. Smart as ever, the animals came to suburban yards for food, with disastrous consequences.

With LA's history in mind and a rapidly growing coyote population approaching 2,000, Chicago by 1999 was witnessing annual removal of three to four hundred "nuisance" coyotes, which seems like a lot until you realize that in the first decade of coyote presence in any city, simply being seen and recognized made a coyote a "nuisance." By 2001 coyotes were so much in the news in the Windy City that in that year Chicago homeowners listed them—not street gangs, not burglars, but coyotes—as the single greatest threat to their safety. At the time there had not been a single aggressive coyote incident in Chicago. With an even larger coyote population now, the figures for nuisance animals have dropped as residents have become more familiar with Chicago as a coyote town.

And the vast majority of the Chicago coyote population consists of upstanding citizen animals. In a recent study there, only 5 coyotes out of 175 tracked became actual nuisance animals, stalking pets or refusing to back away from people. In fact, of 260 animals radio-tracked in Chicago and Los Angeles two decades after the bite-athon of the 1980s, not one showed any aggression toward people. In the same years, between 2,000 and 3,000 people were bitten by dogs in Chicago alone.

"If anything," Stan Gehrt told me, "some people and some communities in greater Chicago have by now become too accepting of coyotes, tolerating behavior I wish they wouldn't." Biologists like Stan and Denver's Stewart Breck want to make sure coyotes in town are like coyotes in Yellowstone, still wild, still a little nervous about us.

Stewart Breck, a dark-haired, bespectacled, genial PhD who works for the National Wildlife Research Center (a research arm of Wildlife Services), and I talked right after he and colleagues had put on a

Denver urban coyote symposium in December 2014. Breck gave me the impression that Denver's story is somewhat distinct from that of either Los Angeles or Chicago. "Coyotes weren't really a presence in Denver until the 1980s," he told me, but at present the estimate ("conservative by maybe 20 percent") puts 112 coyote packs in the Denver Metro Area, with a summer population of 1,004 animals. A remarkable 90 percent of Denver residents in one recent survey indicated having seen a coyote near their home. That, together with the media stories, makes the Denver public "highly attentive" to coyotes.

Human encounters with coyotes, says Breck, ultimately are both "emotional and political." The broad, lethal removal approach that characterized LA in the 1980s is today favored only by about 12 percent of Denver residents. That same recent survey found that fully 45 percent of Denverites—most of them in liberal, well-to-do suburbs—are actually coyote advocates who want them left unmolested in the city. Even attempts to get Coloradans to haze urban coyotes to keep them uneasy around humans have met with

Coyote on Portland MAX light-rail train.
Courtesy Google Images.

resistance in these suburbs and in places like the university town of Boulder. In liberal twenty-first-century America, identification with predators, with reintroduced wolves in the rural West, and now with coyotes in town is a potent political force in a way never seen before in American history.

But Breck hypothesizes that an urban coyote culture is developing in Denver, one reflective of the brashness and boldness of urban human culture, and that such an urban canine culture is likely to emerge over time in other cities as well. The inevitable outcome at this moment, he believes, is more human-coyote conflicts. While the shyer individual coyotes may survive the gauntlet of rural life or be able to coexist with gray wolves in a park like Yellowstone, Breck argues that in urban settings, avoiding humans or gray wolves has become a nonissue for coyotes. In another mirror (or maybe it's a stereotype) of human urban life, bolder, more novelty-seeking individuals among wild coyotes may be most attracted to cities in the first place. They're the first to become accustomed to thronging humans and sensory overload, the first to take risks and try new things, and, Breck reasons, they're transferring both their genes and urban cultural norms to their offspring.

This urban coyote culture, Breck believes, explains a sharp uptick between 2005 and 2010 of "a lot more aggression toward pets and people" in Denver. Now that the Mile-High City is a coyote town, its human residents may need to be a little less tolerant and laid-back about their junior wolves. Or so Stewart Breck and his associates argue.

Maybe tolerance in the form of an appreciative attitude toward urban coyotes actually means accepting, most of all, a new and different definition of what city life is supposed to be. The notion of cities as outside nature is an old fantasy of ours, but in North America coyotes have entirely undermined it, a fact modern journalists and bloggers in particular need to get real about. In 2011 two Canadian researchers, analyzing 453 articles portraying interactions between people and urban coyotes in Canada between 1998 and 2010, found a stunningly uninformed media coloring what the public thought about those interactions. The stories consistently portrayed coyotes as invaders, a plague, an infestation, as unnatural in cities. Most recently the descriptions have tended toward language depicting criminal behavior, describing coyotes as assailants, brazen bandits, suspects, fugitives, kidnappers, robbers, and lurkers, as depraved and (my favorite) "without souls." Almost half the articles discussed "attacks" on people, even though only three people a year were bitten by coyotes in Canada between 1995 and 2010, during which time that country averaged 300,000 dog bites annually.

The arrival of coyotes in a metropolis initially frightens residents. People tend to react to wild coyotes among them as if they are encountering escaped exotic animals from local zoos, often the only explanation that comes to mind. Once they discover that coyotes are not escapees but an urban-adapted population of wild predators, initial fear translates into specific concerns, disease often central among them. To be sure, coyotes historically have suffered from a variety of ordinary canid pathogens, including canine parvovirus, heartworms, distemper (a viral relative of measles in humans), herpesvirus (again, similar to the primate version), and adenovirus. Sarcoptic mange is another affliction among them, although coyotes suffer from a version of the malady introduced among them by veterinarians in the early twentieth century. The disease that city dwellers new to coyotes most commonly fear is rabies. But unlike foxes, coyotes, except for a small population in South Texas, are not carriers of rabies, although (like us) an individual coyote can be infected if bitten by a rabid skunk, fox, or raccoon.

If coyote colonization in other American cities can teach Boston or New York or Washington, DC, anything, it's to learn everything possible about living with the animals, then kick back, be cool, and enjoy them. Their arrival among us is not momentary. Coyotes were here long before we were, and they're not going away.

After two hundred miles and two weeks by rubber raft through the Grand Canyon, in the fall of 2014 my friend Marcus Buck, a Navajo boatman from the town of Bluff, in southeastern Utah, told me this coyote story.

Among the Indian and white homes in and around Bluff, a particular coyote had been causing trouble, killing sheep, chasing pets, and attracting attention to itself. Bluff is no Chicago, with a population of about 350 people rather than 9 million, but like so many big cities across America nowadays, small towns—like the one where my parents lived in Louisiana—have also acquired resident coyotes. An occasional one becomes a problem. So the Navajo chapter leadership asked Marcus if he couldn't hunt down this coyote and take it out.

A few mornings later Marcus was in his truck on the edge of town, bouncing slowly along a dirt road, his rifle in the gun rack behind his head, when out of the waist-high sagebrush a coyote stepped into the road no more than twenty-five feet away. Marcus braked to a stop. The dust from the truck tires rose into the air, briefly obscured the coyote, then settled. The coyote was still there, standing broadside to the truck. Marcus reached behind him and grasped his rifle.

As Marcus told me this story, we were floating through a calm but gorgeous stretch of the Colorado River below Lava Falls Rapid. Crenulated black lava flows and irregular dikes and blobs of lava decorated both banks of the river. Marcus and I were alone on his raft, the rest of our group bobbing along in five more yellow craft some distance downstream. I was sitting behind him and couldn't see his face as he told me what happened then.

"This coyote walks nonchalantly right in front of me in the road. Looks at me. Sees the rifle. Then you know what he does? He yawns, right in my face! Kind of stretches! Then he turns to look back where he'd come from, and I hear coyotes howl a ways back. So he raises his nose, throws back his head, and howls back at them. I've got my hand on the stock of the rifle, but I still haven't pulled it down. Now he stops howling, turns back to me, looks straight as can be right at me for probably half a minute. Then he strolls casually across the road, in no hurry at all, as if he knew something about me."

"So you never shot." I'd known how this was going to go.

"No, I never even got the gun down. He was just too damned nonchalant, too confident. Something. And I didn't know for sure if it was the coyote everybody was looking for. But you know something? Even if it was, I wouldn't have shot. That coyote was so . . . cool looking. So perfect. He was way too pretty to shoot."

I nodded. Marcus had experienced one of those moments of sympathy with the world, much the way Adolph Murie had in Yellowstone so many decades before when he'd watched a coyote trotting along a trail, tossing a sprig of sagebrush into the air and catching it in its mouth again and again. These are moments of identification, animal to animal. They are rare. Sometimes a moment like that becomes unforgettable, because the dialogue of body language isn't getting filtered by the cultural thoughts in our heads, by loaded language like "depraved" or "gang banger" or "archpredator of our time." Each sees the other for who he is.

Although rare, these are the kinds of moments millions of us need to have so we can coexist with coyotes, urban and otherwise.

CHAPTER 7

Coyote America

In the contemporary Western world, you do not grow up imagining death by predation. A dark-haired, soulful girl from Toronto would have outgrown fears like that by the time she was five. Just turned nineteen, Taylor Mitchell had entirely different things on her mind in late October 2009. A songwriter and performer since age sixteen, she had just released her debut album, *For Your Consideration*, which had landed her a gig at the Winnipeg Folk Festival that summer. On a promotional tour that fall, she decided to take a break with a quick hike in Cape Breton Highlands National Park in Nova Scotia. It was the unluckiest decision of her life, the unluckiest decision anyone anywhere in North America made on October 28, 2009. The next morning she lay dead in the Halifax Hospital from massive blood loss as a result of an attack by two coyotes. Rescuers had found the forty-two-pound male standing over her prostrate form, growling, unwilling to leave the scene. She had been bitten all over her body.

Half a decade later we are still trying to figure out exactly how an alluring young woman with a promising career became the first adult in recorded North American history to die from a coyote attack. Coyotes are fully capable of bringing down animals of one hundred pounds or more, certainly deer, and rarely even larger ungulates. But a human being? As with wolves, coyote cultural training does not include humans as part of the species' prey template. Predator biologists have puzzled ever since over what might have gone wrong that October morning along the Atlantic shore. Did the coyotes confuse a young woman for a deer? Did she yield ground or attempt to run, exciting their pursuit instincts? With so many people in North America and so many coyotes—more of both now than at any previous time in history—did the extremely remote mathematical possibilities of a predatory attack by coyotes on an adult human finally produce that one-in-a-million outcome?

Or was something else going on? In November 2014, in an incident actually several orders of magnitude more common than a wolf or coyote attacking a human, a pack of dogs attacked and killed a forty-year-old woman on the Wind River Indian Reservation in Wyoming. That might seem of only glancing relevance to the Nova Scotia attack, except that some biologists have wondered if Taylor Mitchell's fate in some way was tied to a phenomenon researchers had been aware of for at least two decades, as coyotes appeared in state after state in the East and South. Was hybridization between coyotes, wolves, and domestic dogs, in other words, producing a new canid east of the Mississippi, an animal larger than a western coyote, smarter and more clever than a wolf, with a feral dog's aggressive potential?

That's how the larger story is playing out in popular culture. Two years after Taylor Mitchell died, in 2011, a National Geographic Wild television documentary, *Killed by Coyotes*, told her story on national TV. This was followed in 2014 by "Meet the Coywolf," a PBS *Nature* episode that left its international audience with the impression that the bulk of the eastern coyote population consists of "coywolves," a hybrid wolf-coyote created by human actions. Cape Breton Park in Nova Scotia and Taylor Mitchell's death served as the lead to this show. Even reality TV is weighing in on the larger point. Late in 2014 the Discovery Channel aired a program titled *Beasts of the Bayou* wherein its Venice, Louisiana, protagonists searched for what they referred to as a *rougarou*, a "mutant canine," which the show's narration repeatedly called "an aggressive hybrid mutant wolf" that was "more wild and more aggressive than anything seen before." It took a few minutes before my Louisiana upbringing finally untangled *rougarou* as a garbled *loup-garou*, French for "werewolf." The animals this breathless tabloid TV show found in the swamps? A pack of coyotes that may have included a wolf-coyote hybrid.

This "new" hybridization story is older than we think. Its origins, and causes, very likely go back hundreds and probably hundreds of thousands of years. The remote tangle of canid evolution is the most likely sourcebook for our answers, but the popular media are also right: the history of the human-canid relationship is also a driver of the hybrid phenomenon, just as it drives our continuing warfare against coyotes today. And yet—this is the part I like best—the fact that we too are hybrids, that once again the coyote story is echoing the evolutionary trajectory that made us such a wildly successful species ourselves, is the most delicious irony of all.

Another memory, vivid like I'm looking back into time through polished glass, this one destined to color me for life.

I am fourteen, just a couple of years along from having watched that first *Walt Disney Presents* episode on western coyotes, which aired in 1961. In the woods and among the bayous of Louisiana, the West and coyotes seem very far away, or at least they do when I leave home on this summer morning of 1963 and ride my bicycle a couple of miles down a dirt road to a clearing in the woods I know about. This open glade in the grown-up Louisiana forest is my destination because here I can climb a tree and see fifty or seventy-five yards through the trees and prepare to try out an item I'd found in a hardware store a couple of days earlier.

"It's a predator call," the clerk had told me. "We just got them in. Blow on it, and it sounds like a rabbit in distress. You could call up a gray fox with it!" A gray fox? I'd seen one hit by a car along Louisiana

Highway 1, a beautiful little animal. Not a coyote, but as close as 1960s Louisiana was going to get to a wild predator, I'd figured. So, at mid-morning on this sweltering southern summer day, I crawl into somebody's half-collapsed deer stand on the edge of my clearing, wait for silence to envelop the surrounding woodlands, then try an experimental bleat on a wooden call that had cost three bucks.

The interval between the dying out of the rabbit wail and the appearance of an animal on the edge of the clearing opposite me could not have been more than ten seconds. Later, feverishly replaying every detail in my head, I would realize that I had somehow managed to climb a tree with a rabbit call while a very large wild predator was either hunting or reclining for the day only seventy-five yards away. It was a stunning stroke of luck. But in the moment, with this unexpectedly large creature emerging from the woods at a very interested and purposeful trot, heading directly for me, my head was entirely empty of thought. I was a vessel shot from head to toe with adrenalin.

Pointed ears up, gaze fixed, the animal was coming straight as an arrow's flight for the base of the tree I was in. About thirty yards away, where mid-morning sunlight dappled the clearing in light and shadow, it stopped, peered intently in my direction for a few seconds, then resumed its approach, now in a slow, more cautious walk. I had gotten out of the car and examined that highway-kill gray fox; this was no gray fox. Even through the veil of excitement, I registered clear visual impressions. Sharply pointed ears. A long snout. Yellowish and reddish fur that rippled in the sunlight as it approached. Eyes that appeared almost orange. Most of all—the impression I never forgot—it was coming toward me on remarkably long legs, like it was walking on stilts. I had a dog at home, a shepherd mix that I knew weighed fifty pounds. This creature was easily that heavy, I guessed, but it seemed much taller. And now it was only twenty-five feet away.

Every second from the moment I'd put the wooden call to my lips had seemed to transpire in slow motion, as if we were underwater. But in an instant something broke the spell. Probably a downdraft swirl of air sent it a whiff of me, but curiosity on that canine face changed to alarm faster than I could track it. Next I knew it was loping back the way it had come, its head turned backward toward me the whole time, scanning the forest for the source of the danger. In about the same ten seconds it had taken to appear, it disappeared, leaving me sitting in a tree, shaking like a leaf. The whole episode had unfolded across less than two minutes.

What had I seen? It was no fox, but what other than a fox could it have been? When I sat down in my room that night and wrote the Louisiana Department of Wildlife Fisheries in Baton Rouge, I could imagine only one answer: from twenty-five feet away, I had seen a wolf in the northwest of the state, I told them. Two weeks later I got a reply. It is possible you saw a wolf, someone from the office wrote. We think there are still a few red wolves in the state. But (my official letterhead reply stated) it's far more likely that you saw a coyote, since coyotes are now colonizing Louisiana.

A coyote? Wait, what? A coyote like in the Walt Disney films, an animal of the deserts and of the West? I may as well have rounded a corner and run into a moose in downtown New Orleans. I knew about coyotes, but how could there be a coyote anywhere within two hundred miles of Louisiana? Yet in the early 1960s, they were there, and it would not be too long before I saw one, then another and another. That first wild canid I ever laid eyes on, though, stalking my tree on stilt-like legs while I watched with wide eyes, was a harbinger of a different kind of canine future for eastern America.

In Yellowstone National Park in September 2013, I first began to realize how political the debates about wild American canids have become in the modern era. As it turns out, some of our most iconic twenty-first-century battles about charismatic endangered species surge and ebb very specifically around the coyote.

On our second day in the Lamar Valley that September in 2013, Sara and I had met a pair of California scientists, watching the wolves alongside us, who turned out to be Blaire Van Valkenburgh, of the University of California at Los Angeles's Department of Ecology and Evolutionary Biology, and Robert Wayne, the UCLA molecular biologist and Canidae specialist who has done some of the preeminent genetic work on dogs, wolves, and coyotes. A slight, dark-haired scientist who looked to be in his fifties, Wayne was at the Zoological Institute in London when he started researching the molecular evolution of the canids. The work took him head-on into a very big political controversy. So he hadn't missed a beat when I casually mentioned the obvious resemblance between coyotes and the red wolves of the South where I grew up. "There's a good reason for that," he'd frowned. "Red wolves aren't a true species. They're a hybrid, and more coyote than anything else."

"And he should know," Blaire Van Valkenburgh had added brightly. "He did the genetic work demonstrating that the red wolf is really a coyote hybrid."

The genetic markers Wayne found indicated, as he read the evidence, that coyotes are a kind of wolf that shared a common ancestor with gray wolves down to about 3.2 million years ago, when coyote and gray wolf ancestors began to separate, first physically, then, as distance increased, genetically. The question of that evolutionary relationship has animated US endangered species programs going back to the 1960s and has come to spin prominently around a very specific and unusual animal, which since the time of John James Audubon we have called *Canis rufus*.

Since it entered written history (with William Bartram's *Travels* in 1791), the red wolf has been associated with the Deep South. Its range extended as far west as the Hill Country of Central Texas, however—where coyotes also ranged. In true wolf fashion there were black variations (Bartram named the red wolves he saw *Canis niger*), but more commonly the animal was a cinnamon-buff color. Like coyotes. Although it lived in swamps and deep woods, its remarkably long and spindly legs struck many observers as belonging to an animal designed to course after running prey in open country. Its curiosity aroused, it could stand upright on those long legs. Like coyotes,

John Woodhouse Audubon, *Red Texas Wolf* (*Canis rufus*).
Courtesy Google Art Project.

it could live among people, yet it was quite easy to trap or poison. At forty to seventy pounds, it was bigger than a coyote but smaller than a gray wolf. Naturalist Vernon Bailey, on hearing red wolves howl, wrote in 1904, "Their voice is a compromise between that of the lobo . . . and the howl of the coyote. It suggests the coyote much more than the lobo."

In truth, so much about the red wolf seemed coyote-like that as early as 1962—the year before that fourteen-year-old version of me saw a large coyote-like canid from a deer stand in Louisiana—biologists began to question whether the red wolf was a legitimate species or a hybrid resulting from mating between wolves and coyotes. The US Fish and Wildlife Service, however, designated the red wolf an endangered species in 1967 and later followed the lead of a young biologist named Ronald Nowak, who wrote his dissertation on red wolves. Nowak would go on to become the service's endangered species coordinator and to write a major work on canid evolution, *North American Quaternary Canines*, in 1979 (he's also author of the most recent edition of the Bible-like *Walker's Mammals of the World*, the go-to reference for mammals). For decades Nowak deflected all challenges to the red wolf's legitimacy as a distinctive species with full protection under the Endangered Species Act of 1973. Since then the red wolf story has produced endless controversy, and for one primary reason: coyotes. Coyotes, in fact, seemed to be garbling everything we thought we knew about wolves from the Great Lakes to the Deep South.

Partly in an effort to quell the mounting debate about its wolf-management policies—essentially whether, with the red wolf, it had spent millions recovering an animal that may only have come into being in recent history—in 2012 the Fish and Wildlife Service published a peer-reviewed seventy-five-page monograph that offered up an entire rethinking of American wolves. *An Account of the Taxonomy of North American Wolves* concluded that North America's gray wolves were indeed Asian in origin and had arrived here in at least three different waves only 20,000 or so years ago. Using both classic morphology and the host of new genetic studies, the study's authors shrank the number of gray wolf subspecies from the twenty-three Stanley Young and E. A. Goldman had designated in the 1940s down to only four. They extended the former range of the red wolf up the Eastern Seaboard to Canada. And they reclassified the wolf of the eastern Great Lakes—long a gray wolf subspecies—as its own species, *Canis lycaon*.

Most importantly of all, the study claimed that the wolves of eastern America were ancient continental wolves that had come from the same evolutionary line that produced coyotes: "Coyotes, *C. rufus*, and *C. lycaon* are modern representatives of a major and diverse clade that evolved within North America." While *An Account of the Taxonomy of North American Wolves* conceded that both eastern and red wolves may have "anciently" hybridized with coyotes, it posited that "pure" red wolves *had still existed* in the twentieth century.

Outside the hallways of the US Fish and Wildlife Service, with its commitment to red wolf recovery that predates modern genetics research, the role of the coyote in the red wolf story has struck some scientists as far more problematic than federal policies have acknowledged. The American Museum of Natural History warily skirts the issue, its 2009 monograph on canid fossils leaving off any investigation of red wolves. I asked Xiaoming Wang, of the Los Angeles Museum of Natural History and one of the authors of this study, about that. He responded that the red wolf's omission was deliberate: "As an animal with a mixed morphology, including its fossil relatives, it is not always easy to place with certainty. Works on ancient DNA may have a better chance of resolving the mystery."

Robert Wayne and his colleagues have been approaching the debate by looking not at fossil DNA but at the comparative genetics of existing animals. In 2011 Wayne and no fewer than eighteen coauthors published the results of their most recent work on the topic, an article titled "A Genome-Wide Perspective on the Evolutionary History of Enigmatic Wolf-Like Canids." It directly challenges the conclusions and policy of the Fish and Wildlife Service. In the most thorough genetic study so far, the researchers tested genetic markers from 208 gray wolves, 12 red wolves, and 57 coyotes. In contrast to Fish and Wildlife's conclusions, Wayne found no evidence that eastern and red wolves share a common lineage. Great Lakes wolves, he insists, are essentially a population of gray wolves with about 20 percent coyote admixture, and that admixture goes up to as much as 40 percent coyote in the wolf population of Ontario's Algonquin Provincial Park. The hybridization events this study pinpoints are decidedly not the result of today's coyote manifest destiny. Using molecular clock dating strategies, the researchers determined that coyote hybridization with gray wolves in the Great Lakes region initially began between 550 and 950 years ago.

As for the red wolves the federal government has lavished so much effort on recovering? "Structure analysis consistently assigned 80 percent of the red wolf genome to coyotes," the geneticists argued. Large blocks of coyote DNA in red wolf populations imply to Wayne and his colleagues that the creature we now call the "red wolf" is not an ancient American wolf surviving into our time but a hybrid that originated when coyotes began to interbreed with southern gray wolves at some point between 290 and 430 years ago. "We find a coyote-wolf admixture zone that stretched from southern Texas to the Great Lakes and the northeastern U.S. This admixture zone is the largest in area ever described for a terrestrial vertebrate."

Wayne links the dates for these hybridization events to humans, specifically the arrival of Europeans and the changes that followed them. But it doesn't take much of a historian to realize that virtually none of the geneticists' molecular clock dates could possibly reference European history in the Americas. No Europeans were altering landscapes or killing wolves in the Great Lakes area in the 1460s, let alone in 1060. As for the South, only Wayne's event date of 290 years in the past (i.e., 1721) allows for any European influence at all. The southern colonies with red wolves closest to a potential coyote invasion would have been Louisiana and Texas. Louisiana was not settled by the French until 1714 (in Natchitoches) and 1718 (New Orleans), and Texas did not acquire a permanent settlement until Spain founded San Antonio in 1718. Any appreciable European effect on predators in those areas wouldn't have occurred until decades after 1721 from humble beginnings like these.

However, one series of events in the South could provide a historical story to support the time frame in Wayne's molecular clock. The actors were not Europeans, though, except indirectly. Indians had invented agriculture long before Europeans arrived. Native crop growing reached its heyday in the South in the form of the so-called mound-building Mississippian Culture 1,000 years ago. A large agricultural Indian population that also regularly burned the woods subsequently transformed the South, creating a more open setting that featured bluestem-grass prairies in many areas.

Then around 1500, two additional transformations took place. The first European explorers traveled through dense Indian settlements, and in their wake Old World diseases unknown in the Americas swept through the Indian villages, killing millions. That occurred simultaneously with another change, the onset of the three-century weather anomaly known as the Little Ice Age. It produced cooler, wetter conditions that grew such bumper crops of grasses on the Great Plains that swollen bison herds began colonizing in all directions, including into the South. Although not a single early European had seen buffalo in the South earlier, beginning around 1650 and lasting until at least 1725, European travelers reported significant numbers of bison along the woodland trails and in the bluestem prairies of the Deep South.

Early in the 1700s the Indian population began to build up again, European settlers arrived, and the bison herds began falling back westward.

I suspect Wayne's coyote–red wolf hybridization in the South 290 to 430 years ago refers to this bison event. Coyotes must have followed bison from the plains into the Southeast 350 years ago, and over their seventy-five-year stay, a good number of stilt-legged red wolves and sharp-nosed coyotes formed pair-bonds and had litters. By the time coyotes retreated west with buffalo, they'd left a genetic imprint on the canids of the Southeast. The red wolf had became the coyote-like wolf the naturalists would describe a hundred years later. Perhaps some different but corresponding Indian-bison event of 550 to 950 years ago took coyotes into the Great Lakes country and eastward, with similar results.

I did not know until many decades later that as an adolescent growing up in Louisiana in the early 1960s, I had been downstream of a biological tsunami sweeping on padded feet in the direction of the Mississippi River. I had gotten a confusing glimpse of a once-in-several-hundred-years phenomenon. It was sort of like being there to witness the 1883 eruption of Krakatoa but not grasping the bigger picture of plate tectonics. What some canid biologists now describe as one of the most extensive "hybrid swarms" of animals in the North American historical record began moving eastward from the edge of the Great Plains in the late 1940s or early 1950s. I got to witness it.

The "hybrid swarm" in question consisted of wild canids that were partly coyote and partly wolf. The wolf was of course the red wolf. As for the geographic source of the hybridization act, it appears to have been the very Edwards Plateau where proud Texans had all but eradicated predators in the decades from the 1920s through the 1940s. This Texas Hill Country had long been famous for preserving an interesting and unusual intermixed ecology. The hilltops there were grassy and studded with cactus and junipers typical of the edge of the Great Plains. But the river valleys presented a relict southern woodlands setting that even included bald cypress trees.

No one knows just how long western coyotes and southeastern red wolves had ranged through this intermingled West-South setting. But in the 1920s the Biological Survey, with typical fervor for the job, killed 2,209 red wolves in the Texas-Oklahoma-Arkansas region. When Texas's version of Animal Damage Control then subjected the Hill Country to a predator scorched-earth campaign, the pressures led the remaining wolves to hybridize with coyotes. Then, in a move straight out of the main plotline of the coyote's twentieth-century story, unceasing human harassment in one place led this

Coyote–red wolf hybrid, Louisiana, 1960s.
Courtesy Dan Flores.

"swarm" to colonize another. The direction of least resistance was eastward, where the disappearance of the southern wolf population was creating a vacuum that hybrids and coyotes rushed to fill.

By the time I was sitting in my tree, open-mouthed at the leggy canid below me, biologists were daily becoming more aware of the eastward advance of hordes of hybrids and coyotes across the mid-South. Ron Nowak, growing up in New Orleans, was still a decade from writing his dissertation and becoming the Fish and Wildlife Service's endangered species guru. But he was already interested in red wolves and became alarmed enough at stories of the coyote invasion that—moved in part by letters like mine—he actually visited my home parish, Caddo, and other northwestern Louisiana sites in 1965 and 1966, looking for remnant populations of red wolves. In the two years before red wolves were first listed as an endangered species, he found only hybrids and coyotes. As a doctoral student, in 1970 Nowak figured there were no more than three hundred true red wolves left wild anywhere on the continent, with genetic swamping as a result of advancing coyotes and hybrids now the primary threat to their existence.

In the lead-up to passing the Endangered Species Act of 1973, biologists trying to protect red wolves as a species dreamed up some wildly impractical schemes to "save" them from coyotes. One that ought to resonate in our age was a canid-proof fence they hoped to build north-south through East Texas and Oklahoma! When that seemed too daft to implement, the Fish and Wildlife Service brainstormed a canid-free "buffer zone" reminiscent of the infamous "fur desert" the Hudson's Bay Company had attempted to create 150 years earlier to keep American trappers out of the Rockies. Biologists actually planned to kill every coyote and hybrid entering the buffer zone in order to protect the purity of red wolves beyond the line.

In truth, without a sound grasp of canid evolution or any good genetic science to back up their hunches, 1970s biologists working on the coyote-wolf hybridization phenomenon in the South (and, in the same decades, far to the north in Algonquin Provincial Park in eastern Canada) were mostly shooting from the hip. When the

Endangered Species Act became law in 1973, Fish and Wildlife appointed Curtis Carley the first field coordinator for the Red Wolf Recovery Program. Working primarily in southeastern Texas and heroically trying to untangle, among the canids he trapped, which was a wolf, which was a hybrid, and which was a coyote, Carley developed a technique using morphology measurements and recorded howl profiles. It wasn't quite *Crania Americana*, Samuel Morton's mid-nineteenth-century treatise on his human-skull studies using buckshot to determine cranial capacity for the various "races," but it was close enough.

Carley decided early in the project that there was only one possible way of saving red wolves from genetic swamping by coyotes. Biologists were going to have to capture every red wolf remaining in the wild for placement in a captive breeding program. In effect, preserving the red wolf's purity required first bringing about its extinction in the wild and turning its former range over to coyotes and hybrids until biologists could produce enough "pure" animals, then finding a suitable protected preserve for releasing a captive-bred population into the wild again.

How difficult was that? After establishing a certified captive breeding program for red wolves at Point Defiance Zoo in Tacoma, Washington, in 1974 and 1975, the Red Wolf Recovery team decided to examine as breeding candidates some fifty red wolves held in almost twenty zoos across the country. Using the morphology-howl criteria they had established, out of those fifty they identified but a single red wolf, a female in the Oklahoma City Zoo. They were convinced all the rest, plus their pups, were actually either coyotes or hybrids, and in the latter case the team insisted they be destroyed. When some of the shocked zoo personnel refused such a draconian order, in the name of purity Curtis Carley carried out the death sentences himself.

In 1980 the Fish and Wildlife Service proclaimed the red wolf to be extinct in the wild. Out of some four-hundred-plus canids its trappers had captured in Louisiana and East Texas, the program had recognized only forty-three as real red wolves and designated just

A Louisiana anticoyote program resulted in scenes like this in the 1970s.
Courtesy Dan Flores.

seventeen as breeding candidates, three of which were unable to have pups. So with just 14 animals selected out of at least 450, the United States' federal wildlife agency began to breed red wolves in captivity. All those other captured canids deemed to have been betrayed by the sin of contaminating coyote blood? The team destroyed them, every one.

In the full light of day half a century later, this sounds like a wild canid version of the eugenics movement of the early twentieth century, designed to keep the American population untainted by genetic undesirables, with Curtis Carley recalling Madison Grant, the early-twentieth-century conservationist and eugenicist who urged genetic purity in the name of a larger good. Now the task for this increasingly Dr. Strangelove–like program became finding suitable locations for releasing red wolves—locations where, of course, horny coyotes could not get at them.

In Canada, New England, and the northern states, the battle for canid purity hasn't been fought at the same level of intensity as in the

South, although the new "canis soup" did pretty much squash plans to launch recovery of endangered wolves in the Northeast. Both in ancient times and in the last century, coyotes certainly interbred with the Great Lakes' wolves, whose genome now seems about 20 percent coyote. They've contributed considerably more genetics to the eastern wolves of Algonquin Provincial Park, where mingled packs of wolves, hybrids, and coyotes are more common every day as members of a newly formed "coywolf" population. Farther south, as coyotes have spread across states like North Carolina, it has taken heroic efforts of every kind to beat coyotes back from the jealously guarded, endangered red wolves finally released into the wild in Great Smoky Mountains National Park and the Alligator River National Wildlife Refuge in 1987.

The Red Wolf Recovery Program, to its dismay, has gotten little help from the red wolves themselves in what it has called "hand-to-hand combat against the invading coyotes." Because red wolves so readily mate and create pair-bonds with coyotes, the program had to abandon the national park site in 1998 and pull back its recovery efforts from the Alligator River refuge to the more defendable Albemarle Peninsula of the Atlantic shore. It also—obsessively—had to find and destroy hybrids. According to David Rabon, the current coordinator of Red Wolf Recovery, in the mid-1990s biologists realized to their horror that they had missed by two generations a single hybrid pair-bond between a male coyote and a female red wolf. Following existing protocols would have compelled them to destroy the entire wild red wolf population of nearly one hundred animals. Sanity finally prevailed, but this story puts to the test our own conceits about species "purity" and even what we thought we knew about why hybridization happens at all.

For example, why is it that coyotes have made time with wolves in the Northeast and South, yet we preserve no record, from either history or the present, of coyotes and gray wolves hybridizing in the West? The size difference is the answer, some biologists have told me. Yet Mexican gray wolves are also small, the size of red wolves or Algonquin wolves, but scant or no record of hybridizing between coyotes and Mexican wolves exists. Some suggest that hybridization occurs when humans have driven one of the species to rarity. Yet didn't we extirpate gray wolves in the West a century ago, affording all kinds of last opportunities for wolf-coyote hybridizing? We have records of the last gray wolves mating with dogs but not coyotes.

And if rarity of potential mates really is the sole explanation, why do today's red wolves, which the Red Wolf Recovery team has with much effort maniacally winnowed of coyote genetics, have to be pried loose from coyote mates so often? The North Carolina program has actually had to resort to capture and sterilization of the coyote population bordering the refuge in hopes that sterile "space holder" coyotes will create buffer territories against fertile newcomers that might prove irresistible to red wolves. With something like desperation, biologists have looked for natural "behavioral reproductive barriers" between coyotes and red wolves. They're still looking. Something strange is going on here, and it looks very much as if the answer leads us back to the evolutionary lineage of America's canids. What we are seeing happen before our eyes in the eastern United States appears to be something very much like canid behavioral recognition of evolutionary kinship.

The geneticists are offering us two different explanations of what's happening. In their 2011 paper on enigmatic wolflike canids, my Yellowstone wolf-watching companion Robert Wayne and his co-authors argue that genetics show neither a distinctive red wolf genome nor any proof that wolves from Canada and the South were originally related to one another. Further, they argue, Canadian and southern wolves were all originally *gray* wolves, not distinctive American wolves related through deep evolution to coyotes. Eastern and southern wolves, they insist, are related to coyotes now because of hybridization events, in the deep past and recently.

The other argument, also based in genetics, offers up a different scenario. Its advocates, a group of Canadian researchers led by Paul Wilson, contend they can find no gray wolf mitochondrial DNA in either eastern wolves or red wolves and that those animals have coyote-like genetics not due to ancient hybridization but because they have come out of a purely North American lineage of canids that split from coyotes only some 300,000 years ago. Wolf guru David Mech, who argues that the Great Lakes wolves he's studied are actually gray and eastern wolf hybrids, seconds their ideas.

Mech also points out that killing coyotes, not mating with them, is intrinsic to gray wolf behavior. Julie Young of the Predator Research Facility even told me that in experiments there, coyotes inseminated with gray wolf sperm actually killed the puppies they bore. The Canadian argument thus offers up an evolutionary explanation for why coyotes and eastern-southern wolves are so readily hybridizing in our time. *Canis lycaon* and *Canis rufus* breed and form packs with coyotes because they "recognize" one another. Coyotes and gray wolves do not.

Robert Wayne, however, continues to insist that his own study has settled the debate with its different scenario. After the Fish and Wildlife Service in 2012 followed the outlines of the Canadian argument with its monograph *An Account of the Taxonomy of North American Wolves*, in 2014 Wayne persuaded the agency to halt its wolf programs to take account of the "unresolved science."

Regardless of who ultimately wins this scientific debate, there is a natural and logical solution to the penchant eastern and southern wolves and coyotes have for one another. The animals certainly know what that is. I respect the heroic efforts of so many to preserve the red wolf, but if a single coyote getting lucky with a red wolf—in a North America now blanketed with coyotes—threatens the whole show, the work of red wolf recovery simply seems too intrusive and heavy-handed to be lasting.

Besides, as has happened so often and with so many species in the past, genetic purity may be just a momentary accident of time and geography anyway. Both evolution and logic seem to point to a different solution. Why don't we just let coyotes and red wolves do what coyotes and Algonquin wolves are doing already, what comes naturally to them as bequeathed by the Coyote god of North American canid evolution? Let them form pair-bonds, raise litters, and take joy in their creation of a new (or maybe not so new) American animal better adapted to modern conditions. They'll return wild canids to continental ecology, and we'll take joy in having them among us. We'll eventually sort out what to call them, and with their coyote genes, we won't need to continue protecting them so religiously. Then we can just value them as the remarkable new outcome of North America's long, unique canid history.

In the eastern half of North America, where having coyotes—let alone "coywolves"—loping down city streets still appears to some like a scene from a sci-fi movie, life with a midsize predator was on a lot of minds with the death of Taylor Mitchell in 2009. That brought big-time media attention to coyotes, hybridization, and urban predators. For that matter, so have brand-new terms like "coywolf" and "canis soup," the latter coined by *Scientific American* when it noted that across the past century, the last 75 to 140 generations of coyotes have thoroughly mixed with the East's remnant wolves, plus a few dogs along the way.

One result is that the creation of what some are now calling the "eastern coyote" and others the "coywolf" bears an uncanny resemblance to the legendary melting-pot process that continues to birth Americans out of our multicultural immigrant population. In his famous book *The Passing of the Great Race*, Madison Grant lamented and feared the melting pot. Are we being hypocritical to find sentiments like Grant's about multiculturalism to be old-fashioned and very distasteful now, while at the same time we fear the results of similar mixing among the wild canids around us?

Americans are not even merely a wild mélange of ethnicities. What we are witnessing with wild canids in the eastern United States looks very similar to what took place in Europe 40,000 years ago when two or three kinds of humans encountered one another and saw commonality rather than difference. Those of us from a European background, we've recently discovered, are hybrids of another, far more ancient sort. When our species, *Homo sapiens*, left Africa and finally made it through the Middle East and into Europe some 43,000 years ago, we found the country to the north already inhabited by two distinctive human species, Neanderthals and Denisovans. Especially in the first 5,000 years of initial encounters, it seems, we seduced one another a sufficient number of times that, for us out of European lines, 2 to 3 percent of our genetic heritage today comes from these now extinct hominin types.

Just as coyote genes are lending canid hybrids the ability to thrive amid human density, while their wolf genes are helping them live in forests, hunt deer, and (according to eastern coyote biologist Roland Kays) expand their range in the East five times faster than pure coyotes could have, we've preserved a hybrid genetic legacy ourselves. There is a good chance that the genes for light eyes and the mutations that produced fairer skin tones originated with ancient Neanderthal adaptations to life in cold, gray places. Those genes and others acquired from Neanderthals had a selective advantage in northern latitudes, where vitamin D from sunshine was in short supply, because they helped us survive without succumbing to diseases like rickets.

So hybridization is not so bad a thing, especially for species migrating to new places or dealing with rapidly evolving conditions like climate change. It speeds up adaptations dramatically, producing new creatures well fitted to their environments. Just as coyote-wolf hybrids are preserving the genetics of nearly vanished wolves, we have ourselves preserved 20 percent of the genome of the extinct Neanderthals, dispersed in small snippets throughout global humanity.

In modern Coyote America, coexistence with coyotes is an essential lesson, something we need to make second nature as quickly as we can. Coyotes have been in North America far longer than we, they are not going anywhere, and history demonstrates all too graphically that eradicating them is an impossibility. This is truly an instance in which any desire on our part to control nature is perfectly countered by a profound inability to do so. It's a misunderstanding that is a short road to madness in the classic fashion of *Moby-Dick*. Because with coyotes, as with the great whale, resistance is futile.

I've lived in the urban-wildlands zone—in West Texas, Montana, and now New Mexico—for much of my adult life, surrounded by coyotes in each location. Living among them, I have never felt them a threat of any kind. They've functioned for me more as part of the occasional magic show of life, like whales breaching an ocean surface or the Galilean moons, glimpsed through a telescope, swinging around Jupiter over the course of a starry winter night. The coyote's remarkable resilience doesn't just put me in mind of us; it operates as shorthand for the greatest story ever told, the miracle of ongoing evolutionary adaptation to an endlessly changing world. Coyotes are the perfect expression of how life finds a way. They are also one of the iconic life-forms birthed in our part of the globe, an American original that makes us more American the more we know them. We and they are similar success stories in our shared moment on Earth. That's how I, at least, see the coyotes around me. How they see me, I can't know. But I do know this: when I make eye contact with a coyote, I can see the wheels turning inside her head. If I have a theory of mind, so does she.

―――――

It is the high summer of 2014 in the High Desert of the American Southwest, the area that spawned coyotes a half million or more years ago, and like tens of thousands of other Americans in our present age, I'm having a coyote experience. My two-year-old malamute, Kodi, and I have set out on this summer morning on a ramble into the canyon below our house in the urban-wildlands countryside where we live, seventeen miles outside Santa Fe. Half a mile from the back patio, on our way to a spot where the streambed long ago sliced a narrow defile through a soaring lava-flow dike, the quiet July morning has suddenly turned tumultuous.

It begins with excited, alarmed yelps and barks, then unmistakable coyote howls from very close by and from both sides of the canyon.

To our right, perhaps twenty-five yards away, a skinny year-old coyote emerges from the junipers, barking and howling her head off. It only takes me seconds to spot her companion on our left, a sleek young male probably from the same litter, who's taken up a sentry position about the same distance from us. In another minute, I spot a fully grown adult, from the looks of her a female. She hasn't joined the yapping and howling but is standing still and silent, thirty-five yards away. Her unblinking yellow eyes are locked on Kodi.

Kodi is not leashed, but he is trained to my voice. I bring him to my side, noticing in the cacophony of sound how calmly he obeys. At twenty-two months he weighs nearly 130 pounds, and as malamutes do, he resembles a wolf. The coyotes never take their eyes off him and scarcely even glance my way as I unlimber the camera and snap a few shots. After a few minutes of everyone watching everyone else, Kodi and I decide to forego our further progress down-canyon and turn and start for home. We are escorted for a couple of minutes by the little female, who walks along perhaps twenty-five to thirty feet away, incessantly yapping the whole time.

So here's the thing. Coyotes are a fresh enough presence across much of America that someone new to coyotes or who hasn't yet made an effort to understand them could find this scene alarming.

Yearling coyote guardian at its pack's den.
Courtesy Dan Flores.

The coyotes did not run away! They were close! They followed! Someone might have gone for a rifle or called a local wildlife officer to report "aggressive, problem animals." Coyotes are feared—and removed—all the time for reasons like these.

But these coyotes, which we saw many more times through the summer, were neither dangerous nor problematic. They were reacting entirely properly. They had established a territory in my canyon, and we—especially Kodi, a competitor canine—had violated their boundaries. At another time of year, they would have loped away over the rimrock, but in high summer the female probably had pups a few months old hidden away in the canyon's crevices, with her yearlings from a prior litter serving as sentries. I never saw the pups or the alpha male. But we enjoyed hearing this pack sing and occasionally spotting them hunting mice all the rest of the summer.

They were not a threat, some unmanageable wildness near home to fear and destroy in a battle for civilization. They were, rather, an addition to modern life and a beautiful benefit to living in today's Coyote America.

EPILOGUE

Coyote Consciousness

Since almost everyone in America is now living out modern life in a sea of coyotes and coywolves, coyote stories continue accumulating at a dizzying pace. Chased away from the runways at the Portland airport, a coyote dashes aboard a mass-transit bus, which it apparently intends to take downtown, as an airport employee snaps photos of it curled up across the aisle like a normal seatmate. In Chicago, coyotes take to hanging out around the ticket office at Wrigley Field, "scalping game tickets," one sportswriter speculates. At the Predator Research Facility in Utah, an experiment to test boldness finds that some coyotes will approach and take a treat in less than an hour even with massive "deflection stimuli" in place. The monthlong test ends with several other coyotes never having approached the treat once. In another incident there, a graduate student raising a coyote in a state of domestication has to end the experiment suddenly when the coyote reaches an age where it decides the student is an unworthy alpha and the situation almost overnight becomes too dangerous to continue.

No twenty-first-century coyote story, though, quite matches the head-scratching spectacle of a bit of performance art done in New York in 1974. Over a period of three days at the Rene Block

Gallery, an unlikely pair of performers—German artist Joseph Beuys and a wild coyote said to represent America—enacted what Beuys called "I Like America and America Likes Me." Surviving video of the performance shows that Beuys's assistants transported him on a stretcher from the John F. Kennedy Airport to the gallery via ambulance. Once there, entirely covered by a shapeless felt blanket, proffering the hooked end of a cane, and tossing his gloves into the air in an invitation to catch, he and "America" interact. The coyote, for its part, seems pretty much all-in—active, unafraid, a full participant. At the performance's end, with the United States symbolically healed by this shamanic act, Beuys takes his stretcher and ambulance back to the airport and flies home to West Germany, having never actually placed a foot on US soil or seen anything of the country except the gallery and his winning coartist, "America."

Beuys and "America" may not have cured a wobbling United States of its Vietnam era blues in 1974. But unintentionally, I think, "I like America and America Likes Me" reset on the East Coast an earlier, western artistic tradition featuring a self-aware twentieth-century kind of "coyote consciousness." The original version dated to the West Coast literary scene's discovery in the 1950s of the Old Man Coyote stories, which became a homegrown influence that affected both writers and visual artists in the West for the next half century. Now that coyotes are a national phenomenon, with newspaper cartoons and art featuring urban coyotes on the rise in eastern cities, deity/avatar Coyote has emerged from the West to conquer the East. Coyote consciousness seems poised for a second act.

In the West, coyote consciousness happened this way. Federal Indian policy marginalized American Indians on reservations in the late nineteenth century, but Indian storytellers—increasingly stripped of ceremonies and even their languages—never gave up their Coyote tales. The stories would have remained known only to the country's native population, perhaps destined eventually to die out and be lost, but for the work of ethnographers and folklorists who were working to salvage whatever they could of Indian cultures. George Dorsey of the federal Bureau of Ethnology and writer-folklorists Charles

Lummis and Frank Bird Linderman among them, these reservation visitors assisted in making a literary genre out of the Coyote canon by transcribing the oral stories into written texts during the country's early-twentieth-century fascination with American folklore. As Barry Lopez wrote of this period in his own book of Coyote stories, *Giving Birth to Thunder, Sleeping with His Daughter*, by the 1930s the Coyote genre had turned out to be so vast it came close to swamping the American Folklore Association with a bottomless well of Coyote adventures from scores of tribes.

One small but influential group of people—the poets and hipsters of the emerging California counterculture of the 1950s—read and were enthralled by the transcribed Coyote texts. Two decades after serving as one of Jack Kerouac's slightly fictionalized protagonists in *The Dharma Bums*, famed California poet Gary Snyder preserved in his 1977 book *The Old Ways* some of the history of how Coyote captured the West Coast literary scene.

A man named Jaime de Angulo, Snyder tells us, became the rootstock of Coyotism in West Coast literary circles. Angulo, a Spanish MD and linguistic anthropologist who had worked among the Indians in the Southwest and California, went on to become what Snyder intriguingly calls "a San Francisco and Big Sur post–World War II anarchist-bohemian culture hero."

Deflected from an academic career by a lifestyle that apparently was too bohemian even for California tastes, Angulo moved to a hilltop on the Big Sur coast and spent his time collecting California Indian folk tales, most of them about Coyote, finally publishing a 1950s best seller called *Indian Tales*. The book exposed many of the Bay Area and Central Coast writers—Robinson Jeffers and Henry Miller were two of them—to the deity character Coyote for the first time. Angulo must have been a force of nature. He appears in several of Jeffers's poems as "the Spanish Cowboy," as well as becoming a character in some of Kerouac's books. Psychologist Carl Jung considered Angulo the very personification of West Coast Coyotism.

One result of this Pacific Coast flowering of coyote consciousness was a poetry magazine called *Coyote's Journal*. Founded by James

Koller, a Beat poet and novelist originally from Illinois, *Coyote's Journal* (and later a book-publishing imprint called Coyote Books) initially printed the works of the writers and poets associated with City Lights Bookstore. The story is murky, but Koller seems later to have moved the magazine to Albuquerque and eventually to Maine. *Coyote's Journal* appeared in eight issues between 1964 and 1967, one more in 1971, and a last in 1982. It published West Coast writers like Snyder, Koller himself, Allen Ginsberg, Philip Whalen, Ed Dorn, Richard Brautigan, Peter Coyote, and a host of other poets of what was then called the New American Poetry, a literary renaissance that grew out of the Beat and 1960s counterculture movements. Riffs on Coyote were a favorite theme of the body of work.

What about Coyote so appealed to the counterculture? As Snyder put it, "The first thing that excited me about Coyote tales was the delightful, Dadaistic energy, leaping somehow into a modern frame of reference." Like wild coyotes, Coyote also belonged to these writers' home place, North America, and many of them were thinking of themselves as writing from a coastal vantage about the continent stretching eastward. And in a nonconformist age, Coyote was nonjudgmental. Coyote stories were about human nature; they said virtually nothing about good and evil in the abstract. As Snyder put it, Coyote "presents himself to us as an anti-hero." His character seemed to offer all kinds of creative and nonconforming stimulation for rethinking American exceptionalism and the country's heroic sense of itself, both of which came in for so much reappraisal in the 1960s and afterward. Snyder continued to feature Coyote in his poetry and stories for decades, riffing on modern American foreign policy maybe most outrageously and relevantly in his 1986 prose poem "Coyote Man, Mr. President, & The Gunfighters."

Indian writers especially have never stopped invoking Coyote. Leslie Marmon Silko, N. Scott Momaday, Simon Ortiz, and many other native writers have forged ahead in infusing their novels, short stories, and verse with the spirit of Coyote in every possible form—as a rebellious and hilarious raconteur who is at once cosmopolitan, disarming, persuasive, and seductive, but also whimsical, opinionated, irrational, chaotic, and self-destructive. He is simultaneously both a hero to someone and an antihero to someone else. Ultimately, naturally, Coyote is as elusive as the meaning of life.

As Acoma Pueblo writer Simon Ortiz said in his book of Coyote poems, *A Good Journey*, Coyote after all is "the existential Man, Dostoevsky Coyote." As avatar of both coyotes and humans, the Coyote of literature and art is free of cultural conventions of every kind. Coyote is true only to human-coyote nature. But he is very useful, because Coyote somehow manages to get a free pass to speak truth about humanity. And about America.

One coyote who has been speaking truth for more than a half century has been not only a self-described "supergenius" but also an international ambassador of American culture. For much of the period from 1950 to the end of the century, he effortlessly hijacked coyote consciousness. The world's most famous coyote first appeared on broadcast television in September 1949. Beginning in that year, Wile E., known in the early films simply as "the Coyote," made his entrance in full coyote avatar guise—standing upright, making eye contact with the audience—to become the primary protagonist in a quarter century of Looney Tunes and Merrie Melodies cartoons produced for Warner Brothers Studio in Los Angeles.

The compelling TV character that Wile E. became was a collaborative effort between writer Michael Maltese and animator Chuck Jones. Maltese was a New Yorker who had moved to Los Angeles and become the go-to story man for animator-director Jones. Jones was a native westerner who had grown up in the coyote country of Spokane, Washington, then joined a talented group of animators in LA in 1933. The two began fleshing out their endlessly fallible and gullible hero late in 1948.

The genesis of these two animators' movie star coyote easily could have been Old Man Coyote of the Native American stories. In the circles they moved in, the continent's original deity was attracting widespread attention in the 1940s and 1950s. As developed by Maltese and Jones, even Wile E.'s personality seemed uncannily similar to that of the sometimes buffoonish, antihero Coyote. But all the originators of Wile E.—animation master Chuck Jones, in particular—insisted over the years that the idea for Wile E.'s character traits came exclusively from Mark Twain's historic coyote description in *Roughing It*: "a long, slim, sick and sorry-looking skeleton . . . a living, breathing allegory of Want. He is always hungry." Maltese and Jones made Wile E.'s obsession with the Road Runner one of the most fanatical, single-minded pursuits in all of film history, but in the early cartoons Wile E. ate everything from bugs to tin cans on the roadside. He was a coyote gourmand in the Twain style.

Twain's evocation of the Southwest's wild canines had always had a long reach, and it inspired Maltese to develop not just a starving, salivating coyote but a difficult desert prey based on the greater roadrunner, which is actually a ground-dwelling giant cuckoo native to the Southwest. Coyote and Road Runner then assumed their forms via Jones's remarkable animation magic. Their first encounter came in a theatrical cartoon short titled "The Fast and the Furry-ous" in 1949, followed three years later by "Beep, Beep" and "Going, Going, Gosh!" By that point, a pointy-eared, tail-swishing star had been born.

Chuck Jones initially remained with the franchise for only two-dozen episodes before stepping down in 1964, which ended the "classic" period of the cartoons. Although two dozen more Road Runner and Coyote cartoon shorts appeared across the next decade, brought to the screen by different writers and director-animators, the Maltese-Jones creations later starred in the Saturday morning TV series, *The Road Runner Show*, which aired from 1966 to 1968, and in Chuck Jones's 1979 theatrical movie *The Bugs Bunny/Road Runner Movie*. Jones finally returned for a last trio of Road Runner–Coyote cartoons. Two appeared in 1979 and 1980, and a final episode, "Chariots of Fur," aired in 1994.

It seems slightly outlandish, but given pop culture's reach, the world almost certainly gleaned most of what it knows about coyotes—and the generations from the 1950s to the 1970s decidedly did—from casual, sometimes passionate, often joyously stoned delight with the fantasy world of the Road Runner–Coyote cartoons. We learned to associate the red, canyonated deserts of the Southwest with these junior wolves. In the same decades when real coyotes were crossing the Mississippi River and taking up residence in the forests of the East and already eyeing cities there, movie and TV cartoons had us all convinced that the only natural habitat of the coyote was the rectilinear red rock desert. That may have had as much to do with Chuck Jones's growing affection for New Mexico (one of the Chuck Jones Galleries, owned and run by his estate, was long in Santa Fe) as with Mark Twain's original description. In his autobiography, *Chuck Amuck*, Jones says that the series loosely followed a set of nine rules, and according to "Rule 6," "All action must be confined to the natural environment of the two characters—the Southwest American desert."

As for what Wile E. taught us about coyote natural history, the pseudo-Latin binomials the animators dreamed up for coyotes told us everything we needed to know: *Carnivorous vulgaris* (first three shorts, 1949–1952), *Eatibus anythingus* (1954), *Famishius famishius* (1955), *Famishius vulgaris ingeniusi* (1958). *Desertous-operativus idioticus* and *Overconfidentii vulgaris* (both in 1962) provided hints that the Coyote's character was evolving with the times.

With the 1960s, the Mark Twain–inspired Coyote became a Space Age sophisticate. Wile E. now began to acquire a personality more appropriate to a cartoon protagonist from a culture soon to produce Apollo moon shots and the *White Album*. Maltese and Jones had always intended for the audience to identify with the Coyote, and to make his and the Road Runner's adventures universal, at first there was no dialogue ("Rule 4—No dialogue ever, except Beep-Beep!"). As Chuck Jones described his antihero's appeal years later, "Humiliation and indifference—these are conditions everyone of us finds unbearable—this is why the Coyote when falling is more concerned with the audience's opinion of him than he is with the inevitable result of too much gravity."

Wile E. remained nonverbal in all the early cartoons, but eventually the modern world required a Coyote who could explain how to attain the unattainable. First given voice by genius Mel Blanc in the cartoon short *Operation Rabbit* (1952), in both Bugs Bunny cameos and the Road Runner–Coyote shorts the hero became the self-styled "Wile E. Coyotay, Super Genius," a refined, overeducated, decidedly overconfident canine. Wile E.'s casual self-regard seemed perfectly to mirror that of the smug, cocksure policy mavens of the John F. Kennedy–Lyndon B. Johnson years. As the innocent 1950s gave way to a more problematic decade, television's Coyote became a modernist. So in his endless, myopic pursuit, Wile E. became ever more intoxicated with the disease of the American empire in the 1960s: the technological fix.

As a supergenius inhabiting the flickering blue screen of the time, Wile E. was surfing the high tide of American scientific and economic confidence, from which technological solutions for all our problems seemed to flow from unlimited Space Age scientific cleverness. From nuclear power to birth-control pills, from disappearing away all annoying insects to achieving chemical victory over real coyotes on the range, we could solve everything, all the time. In Wile E.'s case, as we all know from deep memory, the ultimate problem solver was not DuPont or Westinghouse but the "Acme Corporation." And Acme's overconfidence rivaled that of the Coyote. In these classic cartoons the promise of the technological fix took the form of Acme Jet-Propelled Roller Skates, an Acme Batman Outfit, Acme Leg Muscle Vitamins, and the Acme Burmese Tiger Trap, so many mad-genius devices of pursuit that today an online poster of Acme's whizbang technologies (it's the work of Chicago artist Rob Loukotka) totals up more than one hundred of them. In iconic, American style, Wile E. trusted every one of those contraptions naively, optimistically, wonderfully. A corporation offers it for sale? *Then of course it's going to work!*

But, of course, like so many of America's easy fixes, then and now, Acme's technology always delivered unintended, funny, and poignant consequences that no supergenius could ever be expected to

Epilogue: Coyote Consciousness 241

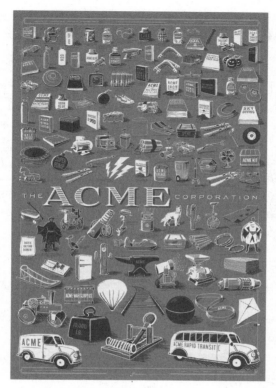

Acme Corporation poster.
Courtesy Rob Loukotka (http://fringefocus.com).

anticipate. We laughed because Wile E.'s plight hit so close to home. Like Old Man Coyote, Wile E. was an exaggerated version of Everyman, and the brilliance in both characters lay in how familiarly they drew their caricatures of human nature. We could identify with everything about them, in Wile E.'s case with his comic overconfidence, his unswerving obsession with a goal, his unfailing faith in technology.

But since Wile E., like Old Man Coyote, was an avatar in a coyote suit, like the continent's original deity the cartoon star spoke to us about ourselves and also about the coyotes of the age. At an important moment, Wile E. gave us the very opposite of the "archpredator of our time." Wile E. was (and is) an entirely sympathetic character.

Wile E. Coyote has pursued the Road Runner in a relentless, televised Möbius strip through the background of all of American life since 1950, until we now reflexively all count a coyote as our friend. He was there again as our avatar when writer Ian Frazier had him star in one of the funniest *New Yorker* stories ever, "Coyote v. Acme," wherein as part of his 1960s hangover, the now litigious-savvy canine determines to bring a product liability lawsuit against the Acme Corporation for flawed technology ("Mr. Coyote seeks compensation for personal injuries, loss of business income, and mental suffering caused as a direct result of the actions and/or gross negligence of said company . . . [inhibiting] his ability to make a living in his profession of predator").

On a couple of occasions, once for a posthumous celebration of Chuck Jones's birthday, I've gone to the Chuck Jones Gallery in Santa Fe and gotten a tour of the inner recesses, where Jones's rollicking, slapstick cartoon creations on the gallery walls—Bugs, Daffy Duck, Elmer Fudd, we know them all—give way to creations he seems to have felt more personal about. My favorite every time I've gone has been an oil painting Jones did that hangs in the gallery manager's office. The first time I saw it, I glanced up from a conversation to witness this thickly rendered swirl of color, the dabs of paint so heavy that they threw shadows, yet in a style at once oddly familiar.

Then I stepped back from the wild brushstrokes and took in the subject entire. It was a self-portrait, but not of Jones. It was Wile E.—tormented coyote face, crenulated nose, cocked and mad eyes, with the head beneath a winter cap swathed and bandaged. I looked closer at the swirling color, the mad glint of the near eye, a little out-of-focus portrait of the Road Runner over his shoulder, in the background, and all of a sudden I saw.

Wile E.'s head was swathed and bandaged because he was missing an ear.

Epilogue: Coyote Consciousness 243

Wile E. Van Gogh, oil on canvas by Chuck Jones, 14" × 9", circa 1992.
Artwork courtesy Chuck Jones Museum. Looney Tunes characters, names, and all related indicia are TM & © Warner Bros. Entertainment, Inc.

Sometimes coyote consciousness has gone for the eroticism at the heart of the human experience, as in Joni Mitchell's 1976 song "Coyote" about her hookup with writer-actor Sam Shepard: "Now he's got a woman at home / He's got another woman down the hall / He seems to want me anyway." The poems in Peter Blue Cloud's *Elderberry Flute Songs*, an imagining of "contemporary Coyote tales" written in 1982, also tend to head off into coyoteroticism. Sometimes coyote consciousness has looked language in the eye, as with Pablo Mitchell's 2005 book *Coyote Nation* and the use of "coyote" as a term, in ever more multicultural America, for all of us whose ethnic backgrounds are blurry. At other times it has presented shopping channel truth—this is America, after all—as when carved, wooden howling coyotes wearing bandannas became fad components of so-called Southwestern style.

With practitioners like painter Harry Fonseca, coyote consciousness has been comic. The Forrest Gump–like Coyote in an entire series of his paintings tends to occupy set-pieces of American history, as in his marvelous piece from a mock Roaring Twenties, *Coyote Does a Rudolph Valentino*. Pay attention and Coyote shows up almost everywhere, sometimes in the most unexpected extraterrestrial places. California science fiction writer Kim Stanley Robinson has a furtive, elusive, dreadlocked character known as Coyote accompany humans

Harry Fonseca, *Coyote Does a Rudolph Valentino*.
Courtesy the Harry Fonseca Trust.

as a stowaway on our next step into the solar system, our colonization of Mars, in his trilogy, *Red Mars, Green Mars,* and *Blue Mars.*

What, in the end, might coyote consciousness grant modern America from the origins of Coyotism in the dimness of continental history? A fallen Unitarian minister named Webster Kitchell may help here. In the 1990s Kitchell published *God's Dog: Conversations with Coyote,* which explores coyote consciousness from just that perspective. Kitchell had come to the Southwest from Amherst and Harvard Divinity School, and his church soon developed a large and loyal following, even after the minister himself suffered a crisis of faith and gave up on God. Ironically, or maybe whimsically, he then turned to Coyote, "a god who was willing to listen sympathetically."

"I met Coyote shortly before dawn on the summer solstice," he writes. "There are coyotes, and there is Coyote. . . . It was at Ghost Ranch, near Santa Fe, New Mexico. I was to meet some people and hike to Chimney Rock to greet the sun on the solstice." Instead, along the way he met a coyote, which ultimately prompted a series of conversations with the American avatar.

"The trouble with you humans," Kitchell has Coyote tell us at the outset of their relationship, "is that weird mind. Somewhere along the evolutionary line you left your animal nature behind. You left behind its truth. You even tell lies about your animal nature, calling it bad or 'lower.'"

As the conversations continue (Coyote often arrives for them driving a red convertible with one or more attractive blondes in tow), the minister upbraids Coyote for having no morality in all the ancient stories about him. "I have an absolute morality," Coyote responds. "I am for me. And a few of my friends." Well, then what about death? Kitchell asks. "Death is a little prod in the back of your awareness which asks if you're hiding from life," Coyote responds. "When you really accept your own death, life becomes tasty and tangible and sensual." Coyote conceded that since he was immortal, his insights about death might not be firsthand, but as far as he could see, mortals had best focus always on life, not hope for something after.

Coyote Cantina, New Mexico, entrance art.
Courtesy Dan Flores.
Picture used with permission of Coyote Cafe, Santa Fe, New Mexico.

By the end of their conversations, Coyote had become a god even a fallen minister could believe in, a god of life rather than death. In Kitchell's view, Coyote even passed Friedrich Nietzsche's test of the superman—to have accepted your life so completely that you would want it to play out again exactly the same way, in every single detail, because it is yours and yours alone.

In our modern world nobody needs another religion, but I do appreciate and enjoy Coyote as social critic and clever philosopher of human nature. So I'm guessing that since the coyote is now a Continental Everyman, Beuys's "I Like America and America Likes Me" may have been the first act in another, contemporary chapter in coyote consciousness, this time on a national stage rather than just a western one. Maybe as a result we'll resurrect the Aztec Coyotlinauatl festivals and all dress up as coyotes once a year. At the very

Epilogue: Coyote Consciousness 247

Coyote tracks in the sand. Coyote is still "going along."
Courtesy Dan Flores.

least a new coyote consciousness should definitely heed that odd line from the first European to write about them when he noted that the coyote was "a persevering revenger of injuries" but also "grateful to those who do well by it." As coyotes go trotting off into the American future, however their story plays out, it will be something to see.

The coyote's biography in North America has always been one of many acts, but in the twenty-first century it is now a fully American story, an adventure from coast to coast. The Hundred Years' War on Coyotes in the American West has certainly pressed on with a very much alive federal killing agency that continues to dispatch a phenomenal number of coyotes every year. In its own clueless way,

Wildlife Services begs the question of how North America ever functioned without us. Its irony as a taxpayer program is that its relentless, lethal harassment of coyotes in the rural West is a principal reason why there are coyotes running through the streets of New York City today. If we actually want fewer of them or want to slow their saturation of the continent, the obvious solution is to stop killing them and allow their populations to stabilize. It's a simple thing, but for a full century now, human nature has been unable to stand back and allow coyote nature to work.

We also ought to value the coyotes trotting through our yards for the avatar stand-in role they play for us. Humanity faces what from all best indications looks to be a noir future, a daunting challenge, environmentally and ecologically. Coyotes have already experienced at least two similarly epic climate swings, one the demise of deep cold and wet, the other a peak of hot and dry. Many, many other creatures did not survive those. Coyotes did, and they originally attracted our attention because of it. They have also survived our own attempt to wipe them off the planet, and we were pretty damned dedicated to that. As our future unreels, I for one am going to be watching coyotes very, very closely to see just what they do.

One thing I already know now. However the world changes, starting from this point, anywhere at all in America, I'll get to step outside at night and hear that half-million-year-old coyote national anthem, sung to the stars and planets as they swing over an old, old continent.

Selected Bibliography

Alexander, Shelley, and Michael Quinn. "Coyote (*Canis latrans*) Interactions with Humans and Pets Reported in the Canadian Print Media, 1905–2010," *Human Dimensions of Wildlife* 16 (2011): 345–359.

Almberg, Emily, et al. "Infectious Diseases in Yellowstone's Canid Community," *Yellowstone Science* 19 (Summer 2011): 16–24.

Audubon, John James. "Missouri River Journal," in *Audubon and His Journals*, Maria Audubon and Elliott Coues, eds. (2 vols.; New York: Dover Press, 2nd edition, 1986): I.

Barrow, Mark. *Nature's Ghosts: Confronting Extinction from the Age of Jefferson to the Age of Ecology* (Chicago: University of Chicago Press, 2009).

Beeland, T. Delene. *The Secret World of Red Wolves: The Fight to Save North America's Other Wolf* (Chapel Hill: University of North Carolina Press, 2013).

Bekoff, Marc, ed. *Coyotes: Biology, Behavior, and Management* (Caldwell, NJ: Blackburn Press, 1978).

———. *The Emotional Lives of Animals* (Novato, CA: New World Library, 2008).

Beuys, Joseph. "I Like America and America Likes Me," performance art video, 1974, https://www.youtube.com/watch? v=e5UXAqpSJDk.

Bird, Isabella. *A Lady's Life in the Rocky Mountains* (Norman: University of Oklahoma Press, 1975).

Blue Cloud, Peter. *Elderberry Flute Song: Contemporary Coyote Tales* (Buffalo, NY: White Pine Press, 2002 edition).

Breck, Stewart. Interview with the author, December 2014, disk copy in author's possession.

Bright, William. *A Coyote Reader* (Berkeley: University of California Press, 1993).

Budiansky, Stephen. *The Covenant of the Wild: Why Animals Chose Domestication* (New Haven, : Yale University Press, 1992).

Carhart, Arthur. "Poisons: The Creeping Killer," *Sports Afield* (November 1959): 56–57, 127–129.

Carriker, Robert. *Father Peter John DeSmet: Jesuit in the West* (Norman: University of Oklahoma Press, 1995).

Carson, Rachel. *Silent Spring* (New York: Houghton-Mifflin, 40th anniversary edition, 2002).

Chambers, Steven, et al., *An Account of the Taxonomy of North American Wolves from Morphological and Genetic Analyses*. North American Fauna 77 (Washington: United States Department of Interior Fish and Wildlife Service, 2012).

Clark, Ella. *Indian Legends from the Northern Rockies* (Norman: University of Oklahoma Press, 1966).

Connolly, Guy. "The Effects of Control on Coyote Populations: Another Look," *Symposium Proceedings—Coyotes in the Southwest: A Compendium of Our Knowledge* (Lincoln, NE: Wildlife Damage Management, 1995): 23–29.

Crabtree, Robert, and Jennifer Sheldon. "The Ecological Role of Coyotes on Yellowstone's Northern Range," *Yellowstone Science* 7 (Spring 1999): 15–23.

Davis, Mike. *The Ecology of Fear: Los Angeles and the Imagination of Disaster* (New York: Metropolitan Books, 1998).

Dawkins, Richard. *The Selfish Gene* (New York: Oxford University Press, 1989).

Dobie, J. Frank. *The Voice of the Coyote* (Lincoln: University of Nebraska Press, 1961).

Dorsey, George. *The Mythology of the Wichita* (Norman: University of Oklahoma Press, 1995).

Dog Zombie. "Guest Blog: The Hearty Ingredients of Canis Soup," *Scientific American*, December 27, 2011.

Dunlap, Thomas. *Saving America's Wildlife: Ecology and the American Mind, 1850–1990* (Princeton: Princeton University Press, 1988).

———. "'The Coyote Itself': Ecologists and the Value of Predators, 1900–1972," *Environmental Review* 7 (Spring, 1983): 54–70.

Erdoes, Richard, and Alfonso Ortiz, eds. *American Indian Trickster Tales* (New York: Penguin Books, 1998).

"Family CANIDAE—Wolves, Coyote, Dogs, and Foxes," in E. Raymond Hall and Keith Kelson, *The Mammals of North America* (2 vols.; New York: Ronald Press, 1959): II.

Flippen, J. Brooks. *Nixon and the Environment* (Albuquerque: University of New Mexico Press, 2000).

Flores, Dan. *American Serengeti: The Last Big Animals of the Great Plains* (Lawrence: University Press of Kansas, 2016).

Fox, Camilla. Interview with the author, November 2013, disk copy in author's possession.

Frazier, Ian. "Coyote v. Acme," *The New Yorker*, February 26, 1990.

Gabler, Neal. *Walt Disney: The Triumph of the American Imagination* (New York: Vintage Books, 2006).

Gehrt, Stanley. Interview with the author, December 2014, disk copy in author's possession.

———. *Urban Coyote Ecology and Management: The Cook County, Illinois, Coyote Project* (Columbus: Ohio State University Extension Service Bulletin 929, nd).

Gehrt, Stanley, Seth Riley, and Brian Cypher, eds. *Urban Carnivores: Ecology, Conflict, and Conservation* (Baltimore: Johns Hopkins University Press, 2010).

Gese, Eric. Interview with the author, November 2014, disk copy in author's possession.

Glutton-Brock, Juliet. "Aristotle, the Scale of Nature, and Modern Attitudes to Animals," *Social Research* 62 (Fall 1995): 421–440.

Goetzmann, William, et al. *Karl Bodmer's America* (Lincoln: Joslyn Art Museum and University of Nebraska Press, 1984).

Gregg, Josiah. *Commerce of the Prairies*, Max Moorhead, ed. (Norman: University of Oklahoma Press, 1954).

Grimm, David. "In Dogs' Play, Researchers See Honesty and Deceit, Perhaps Something Like Morality," *Washington Post*, May 19, 2014.

Grinnell, Joseph, and Tracy Storer. "Animal Life as an Asset of National Parks," *Science* XLIV (1916): 375–380.

Hall, Jon. "Hounded: Dogs, Humans, and the Rise of the American City" (Ph.D. dissertation in History, University of Montana, Missoula, MT, 2016).

High Plains Films and Fund for Animals, *Killing Coyote* (2000).

Hinton, Joseph, David Rabon, et al. "Red Wolf (Canis Rufus) Recovery: A Review, with Suggestions for Future Research," *Animals* 3 (2013): 722–744.

Hurley, M. A., et al. "Demographic Response of Mule Deer to Experimental Reduction of Coyotes and Mountain Lions in Southeastern Idaho," *Wildlife Monographs* 178 (August 2011): 1–33.

Hyde, Lewis. *Trickster Makes This World: How Disruptive Imagination Creates Culture* (New York: Farrar, Straus and Giroux, 1998).

Jones, Chuck. *Chuck Amuck: The Life and Times of an Animated Cartoonist* (New York: Farrar, Straus and Giroux, 1999).

Jones, Karen. *Wolf Mountains: A History of Wolves Along the Great Divide* (Calgary: University of Calgary Press, 2002).

Kitchell, Webster. *God's Dog: Conversations with Coyote* (Boston: Skinner Books, 1991).

Kluger, Jeffrey. "The Mystery of Animal Grief," *Time* (April 15, 2013): 41–45.

Kruuk, Hans. *Hunter and Hunted: Relationships Between Carnivores and People* (Cambridge: Cambridge University Press, 2002), UK.

Leopold, Aldo. *A Sand County Almanac and Sketches Here and There* (New York: Oxford University Press, 1949).

———. *The River of the Mother of God and Other Essays*, Susan Flader and J. Baird Callicott, eds. (Madison: University of Wisconsin Press, 1991).

Lopez, Barry. *Giving Birth to Thunder, Sleeping with His Daughter: Coyote Builds North America* (New York: Harper-Perennial, 2001).

Lummis, Charles. *Pueblo Indian Folk-Stories* (Lincoln: University of Nebraska Press, 1992).

McIntyre, Rick. Interview with the author, September 2013, disk copy in author's possession.

Marsh, George Perkins. *Man and Nature, or The Earth as Modified by Human Action* (New York: Arno and the New York Times, 1970).

Meachen, Julie, and Joshua Samuels. "Evolution of Coyotes (*Canis latrans*) in Response to the Megafaunal Extinctions," *Proceedings of the National Academy of Sciences* 109 (March 2012): 4191–4196.

Mead, James. *Hunting and Trading on the Great Plains, 1859–1875*, ed. Schuyler Jones (Norman: University of Oklahoma Press, 1986).

Melendez, Theresa. "The Coyote," in Angus Gillespie and Jay Mechling, eds., *American Wildlife in Symbol and Story* (Knoxville: University of Tennessee Press, 1987): 203–233.

Mighetto, Lisa. *Wild Animals and American Environmental Ethics* (Tucson: University of Arizona Press, 1991).
Mills, Enos. *Wild Life on the Rockies* (Lincoln: University of Nebraska Press, 1988).
Moulton, Gary, ed. *The Journals of the Lewis and Clark Expedition, Volume 3: August 25, 1804–April 6, 1805* (Lincoln: University of Nebraska Press, 1987).
———. *The Journals of the Lewis and Clark Expedition, Volume 4: April 7–July 27, 1805* (Lincoln: University of Nebraska Press, 1987).
Murie, Adolph. *Ecology of the Coyote in Yellowstone*. Fauna of the National Parks of the United States No. 4 (Washington: U.S. Department of Interior National Park Service, 1940).
Murie, Olaus. *Food Habits of the Coyote In Jackson Hole, Wyo.* (Washington: United States Department of Agriculture, Bulletin 362, 1935).
———. Papers. Conservation Collection, CONS90, Denver Public Library, Denver, Colorado.
Nature and PBS. *Meet the Coywolf* (2013).
Nixon, Richard. "Special Message to the Congress Outlining the 1972 Environmental Program," February 8, 1972. Online by Gerhard Peters and John T. Woolley, The American Presidency Project. http://www.presidency.ucsb.edu/ws/?pid=3731.
Nowak, Ronald. *North American Quaternary Canis* (Lawrence: University of Kansas Museum of Natural History, 1979).
———. *Walker's Mammals of the World* (2 vols.; Baltimore and London: Johns Hopkins University Press, sixth edition, 1999).
Nuttall, Thomas. *A Journal of Travels into the Arkansas Territory During the Year 1819*, Savoie Lottinville, ed. (Norman: University of Oklahoma Press, 1979).
Osburn, Katherine. "The Navajo at Bosque Redondo: Cooperation, Resistance, and Initiative, 1864–1868," *New Mexico Historical Review* 60 (October 1985): 399–413.
Parkman, Francis. *The Oregon Trail* (New York: New American Library, 1950).
Pavlik, Steve. "Will Big Trotter Reclaim His Place? The Role of the Wolf in Navajo Tradition," *American Indian Culture and Research Journal* 24 (Fall 2000): 1–19.
Peterson, Shannon. *Acting for Endangered Species: The Statutory Ark* (Lawrence: University Press of Kansas, 2002).

Pike, Albert. *Journeys in the Prairie, 1831–1832* (Canyon, TX: Panhandle-Plains Historical Society, 1969).

Powell, John Wesley. *Report on the Lands of the Arid Region of the United States, with a More Detailed Account of the Lands of Utah* (Boston: Harvard Common Press, facsimile edition, 1983).

Robinson, Michael. *Predatory Bureacracy: The Extermination of Wolves and the Transformation of the West* (Boulder: University Press of Colorado, 2005).

Russell, Edmund. *Evolutionary History: Uniting History and Biology to Understand Life on Earth* (New York: Cambridge University Press, 2011).

Ruxton, George Frederick. *Life in the Far West*, Leroy Hafen, ed. (Norman: University of Oklahoma Press, 1950).

Ryden, Hope. *God's Dog: Celebration of the North American Coyote* (New York: Viking, 1979).

Salt Lake Weekly Tribune, "A Journey Through Western Utah and Nevada, South of Grantsville," September 8, 1887.

Sandlos, John. "Savage Fields: Ideology and the War on the North American Coyote," *Capitalism, Nature, Socialism* 9 (June 1998): 41–51.

Sankararaman, Sriram, et al. "The Date of Interbreeding Between Neandertals and Modern Humans," *PLoS Genetics* 8 (October 2012): 1–9.

Schimmoeller-Peiffer, Katrina. *Coyote At Large: Humor in American Nature Writing* (Salt Lake City: University of Utah Press, 2000).

Seton, Ernest Thompson. "Tito: The Story of the Coyote That Learned How," *Scribner's* 28 (August 1900): 1–25.

Sheldon, Jennifer. *Wild Dogs: The Natural History of the Nondomestic Canidae* (Caldwell, NJ: Blackburn Press, 1992).

Shepard, Paul. *The Others: How Animals Made Us Human* (Washington: Island/Shearwater, 1996).

Shivik, John. *The Predator Paradox: Ending the War with Wolves, Bears, Cougars, and Coyotes* (Boston: Beacon Press, 2014).

Smail, Daniel Lord. *On Deep History and the Brain* (Berkeley: University of California Press, 2008).

Smith, Doug. Interview with the author, September 2013, disk copy in author's possession.

———. "Ten Years of Yellowstone Wolves, 1995–2005," *Yellowstone Science* 13 (Winter 2005): 7–33.

Snyder, Gary. "The Incredible Survival of Coyote," in *The Old Ways* (San Francisco: City Lights Books, 1977): 67–93.

Spiro, Jonathan. *Defending the Master Race: Conservation, Eugenics, and the Legacy of Madison Grant* (Burlington: University Press of New England, 2009).

Stroud, Patricia. *Thomas Say: New World Naturalist* (Philadelphia: University of Pennsylvania Press, 1992).

Tedford, Richard, Xiaoming Wang, and Beryl Taylor. *Phylogenetic Systematics of the North American Fossil* Caninae (Carnivora: Canidae) (New York: Bulletin of the American Museum of Natural History, 2009).

Tomer, John S., and Michael J. Brodhead, eds. *A Naturalist in the Indian Territory: The Journals of S. W. Woodhouse, 1849–50* (Norman: University of Oklahoma Press, 1992).

Townsend, John Kirk. *Narrative of a Journey Across the Rocky Mountains to the Columbia River* (Lincoln: University of Nebraska Press, 1978).

Twain, Mark. *Roughing It* (New York: Penguin American Library, 1982).

Tyler, Hamilton. *Pueblo Gods and Myths* (Norman: University of Oklahoma Press, 1964).

Van Nuys, Frank. *Varmints and Victims: Predator Control in the American West* (Lawrence: University Press of Kansas, 2015).

VonHoldt, Bridgett M., et al. "A Genome-Wide Perspective on the Evolutionary History of Enigmatic Wolf-Like Canids," *Genome Research* (Cold Spring Harbor, NY: Cold Spring Harbor Laboratory Press, 2011).

Walt Disney Presents. "The Coyote's Lament," 1961. https://www.youtube.com/watch?v=nnSCOcwgjKs.

Wang, Xiaoming. Interview with the author, October 2013, notes in author's possession.

Way, Jonathan. *Suburban Howls: Tracking the Eastern Coyote in Urban Massachusetts* (Indianapolis: Dog Ear Publishing, 2014).

Wayne, Robert. Interview with the author, September 2013, notes in author's possession.

———. "Molecular Evolution of the Dog Family," *Trends in Genetics* 9 (June 1993): 218–224.

Webb, Walter Prescott. *The Great Plains* (Lincoln: University of Nebraska Press, 1981).

Whittlesey, Lee, and Paul Schullery. "How Many Wolves Were in the Yellowstone Area in the 1870s?" *Yellowstone Science* 19 (Spring 2011): 23–28.

Wildlife Management Institute. *The American Game Policy and Its Development, 1929–1930* (Washington: Wildlife Management Institute, 1930).

Wilson, Edward O. *The Meaning of Human Existence* (New York: Liveright Publishing, 2014).

Wilson, Paul, et al. "DNA Profiles of the Eastern Canadian Wolf and the Red Wolf Provide Evidence for a Common Evolutionary History Independent of the Gray Wolf." *Canadian Journal of Zoology* 78 (2000): 2156–2166.

Wise, Michael. "Killing Montana's Wolves: Stockgrowers, Bounty Bills, and the Uncertain Distinction between Predators and Producers," *Montana, the Magazine of Western History* 63 (Winter 2013): 51–66.

Worster, Donald. *Nature's Economy: A History of Ecological Ideas* (New York: Cambridge University Press, second edition, 1994).

Young, Julie. Interview with the author, November 2014, disk copy in author's possession.

Young, Stanley, and Harley Jackson. *The Clever Coyote* (Lincoln: University of Nebraska Press, 1978).

Index

Abbey, Edward, 179
Absaroka National Forest, 140
Acme Corporation, 240–242
Adams, Charles C., 117, 122–123
adaptation, study of, 117
adaptive success, of coyotes, 5, 35–36, 104–107, 148
aerial gunning, 174–176
agriculture, invention of, 218
Albright, Horace, 138
Algonquin Provincial Park, 224
Alligator River National Wildlife Refuge, 224
altitude-based life zones, 122
American Bison Society, 118
American Game Conference, 125
American Great Plains, 25–26, 53–54
   animal extinctions on, 84, 94
   Little Ice Age and, 219
American Museum of Natural History, 125
American Society of Mammalogists, 122, 145
   on predators, 109, 122–125, 134–135

American West explorers, 54–57, 59–68, 76
de Angulo, Jaime, 235
Animal and Plant Health Inspection Service (APHIS), 170
Animal Communities in Temperate America (Shelford), 117
Animal Damage Control Act (1931), 114, 116, 121, 136
   origin of, 109, 134
   proposed length of, 145
   repeal of, 166
   signing of, 135
   wake of, 140
Animal Damage Control Act (1972), 166
   passage of, 167
Animal Damage Control Bill
   opposition to, 134
   passage of, 135
Animals I Have Known (Seton), 103
animism, 27, 29
APHIS. *See* Animal and Plant Health Inspection Service
archpredator, coyote as, 114, 116, 134, 141, 144

Artemis, 27
Audubon, John James
  Canis rufus, 215
  paintings of, 74–75
  on prairie wolf, 74
Audubon Society, 118
Aztecs
  animal fascination of, 10
  coyotes and, 9–10, 48
  deities of, 10

Bailey, Vernon, 95, 122, 125
  on *Canis rufus*, 215
  on coyotes, 96
  mass-extermination techniques of, 100
Baird, Spencer, 64
balance of nature, 117–118
  Goldman on, 124
  Muir on, 108
  niches and, 119
  predators and, 120
  white Americans and, 139
  See also nature
Barton, Benjamin Smith, 60
Bartram, William, 214
behavioral reproductive barriers, 225
behavioral trait evolution
  in animals, 91
  in coyotes, 107–108, 128
  in humans, 39
Bekoff, Marc, 19, 93
Bering land bridge migration, 32–33
Beuys, Joseph, 234, 246
Big Bend National Park, 148
Bigfoot (gray wolf), 113
biocentrism, 154–155, 161
"A Biotic View of Land" (Leopold), 154
Bird, Isabella, 78
bison
  in deep South, 219
  slaughter, 84, 86
Bison Peak Pack, 132
  See also coyotes

Blackfeet, 25
Blanc, Mel, 240
Blue Mars (Robinson), 245
Bodmer, Karl, 68
  fame of, 66
  paintings of, 63, 65
Boone and Crockett Club, 118
bovine evolution, 6
Brautigan, Richard, 236
Breck, Stewart, 193, 202
  on urban coyotes, 203–204
Brooks Range, 5
Bryce National Park, 148
Buchanan, James, 109
Buck, Marcus, 206
Bureau of Biological Survey, 113, 133
  bait stations, 98–99
  coyote characterization by, 144
  extermination policy of, 135
  funding for, 95–97, 109
  Grinnell, J., and, 123–125
  hunters employed by, 97–98
  justification for, 96–97
  mission statement of, 95
  Murie, O., on, 147
  policies of, 118
  predator control through, 95
  scientist relations with, 144
  success of, 103
  See also Fish and Wildlife Service
Bureau of Ethnology, 234
Burroughs, John, 108

Cain, Stanley, 164
Canidae, 26, 214
  evolution of, 30–31
canids, 120
  ancestral, 3
  evolution of, 29–31, 211, 221
  mating strategies of, 105
  North American lineage of, 225–226
  purity of, 222–224
canines
  morality and, 93
  See also wild canines

Index

Canis anthus, 60
Canis armbrusteri, 33
Canis aureus, 32, 60
Canis chihliensis, 33
Canis dirus, 33
Canis edwardii, 31
 coyotes coexistence with, 32
 fossil evidence of, 32
Canis frustror, 67–68
Canis latrans, 32, 75
 Say on, 58–60
 See also prairie wolf
Canis latrans frustror, 68
Canis latrans orcutti, 34
Canis lepophagus, 30–32
Canis lupus, 26, 33
Canis lycaon (Northeastern wolf), 6, 216, 226
Canis niger, 214
Canis rufus (red wolf), 6, 214, 226
 Audubon rendering of, 215
 Bailey on, 215
 as hybrid, 215–216, 219
 See also red wolf
Canyonlands National Park, 148
Carhart, Arthur, 156
 on poisons, 157–158
Carley, Curtis, 222–223
carrying capacity, 105, 133, 148
Carson, Kit, 50
Carson, Rachel, 156–157
 on poison, 158–159
Carson National Forest, 94
Carter, Jimmy, 170
Carter, Robert, 83
Central America, coyotes in, 4
Chaco City, 10
Chauvet Cave, 33
Chico, the Misunderstood Coyote (Disney), 186
Chihuahuan desert, 4
Chisholm, Jesse, 85
Christianity, 27
Chuck Amuck (Jones), 239
cities, as ecosystems, 193, 204–205

Civilian Conservation Corps, 95
Clark, William, 95, 109
 animals observed by, 54–55
 explorations of, 54
 journals of, 191
Clavijero, Francisco Javier, 58
Clements, Frederick, 117
The Clever Coyote (Young, S.), 147
Climax conditions, 117
Clovis hunting culture, 115
Clovis people, 25
Colville, 25
Commerce of the Prairies (Gregg), 69–70, 73
competition, evolution and, 34
Compound 1080. *See* sodium fluoroacetate
Connolly, Guy, 148
Continental Divide, 119
Control Methods Lab, 124
Corbin, Ben, 92
cosmopolitan species, coyotes as, 5–6
Coues, Elliott, on coyotes, 117, 179
A Country Coyote Goes Hollywood (Disney), 187
Coyoacan (religious coyote cult), 9
Coyote (Native American god), 28
 appeal of, 236
 consciousness, 234, 243–245, 247
 creation myth, 21–22
 death rationale, 45–48
 early mythologies of, 37
 emergence of, 25
 genre, 235
 as human avatar, 37, 237, 248
 human nature myth, 22–24, 42–44
 Jung on, 27
 Kitchell on, 245
 as literary character, 38
 selfishness parable, 44–45
 Snyder on, 236
 story functionality, 41–42
 universality of, 41–42
 See also Old Man America
Coyote, Peter, 184, 236

Coyote, Wile E., 237–240
  identification with, 241–242
  plight of, 241
"Coyote and His Knee" (Wichita),
  22–24
"Coyote and the Frog People"
  (Kalapuya), 44–45
"Coyote and the Shadow People"
  (Nez Perce), 46–48
coyote control, 141, 169
  end of in national parks, 139–140
  evolutionary colonizing mechanisms
    and, 108, 148
  in Glacier National Park, 100–101
  kill count, 147
  livestock industry and, 142–143
  methods of, 88, 145–147
  nonlethal, 175–177
  population effects of, 143
  program results, 223
  sentiment favoring, 138–139
  sheepmen on, 102, 142–143
  sterilization, 175–177
  in Texas Hill Country, 147–148
  during WWII, 145
  Young, J., on, 174–175
  See also predator control
Coyote Extermination Act, 109
Coyote Nation (Mitchell, P.), 243
coyote pelts
  fur trade and, 85
  Lewis on, 85
  Ruxton on, 85
coyote power, Navajos on, 49–50
coyotes, 54, 230, 247
  adaptive success of, 5
  ancestor of, 26–27
  anecdotal stories of, 233–234
  as archpredator, 114, 116, 134,
    141, 144
  Aztecs and, 9–10, 48
  Bailey on, 96
  behavioral trait evolution of,
    107–108, 128
  biased predation evidence of, 136

bounties on, 87–88, 96, 152, 184
Bureau of Biological Survey on,
  144
C. edwardii coexistence with, 32
canine competition with, 191–192
in Central America, 4
changing pronunciation of, 71–73
close encounters, 1–2
colonization behavior of, 7–8, 108
as cosmopolitan, 5–6
Coues on, 117
cultural transmission among,
  106–107
dentition of, 36
eradication of, 115–116, 141
evolution of, 3, 26, 34–35, 105–107,
  132–134, 148
extermination war against, 7, 16, 88,
  144–145, 162
fission-fusion in, 104–106, 108
genetic analysis of, 26–27
Goldman on, 109, 125, 134, 137
gray wolves and, 65–66, 125–126,
  214, 217–218
Grinnell on number slaughtered,
  86
habitat range of, 2–5, 7–8, 61,
  101–102, 107, 196
hearing in, 106
Hernandez on, 57–58
history of, 14
howling melodies, 81–83
humans and, 2–3, 6, 9, 14–15, 27,
  191–194, 204–205, 209
Hyde on, 28
in Indian myth, 116
Indian observations of, 14–15
individuality of, 115, 177–178
introduction of, 3
jackals and, 29, 63–64
lack of sympathy for, 138–139
in Lamar Valley, 132
Leopold on, 139
Lewis on, 55–57
litter size, 7, 105

Index

in Mexico, 4
Mills on, 117
mimicry of, 18
Mitchell, T., attacked by, 209–210, 227
Muir on, 108
in New York City, 11–12
nineteenth-century traveler references to, 190–191
as omnivorous generalists, 35, 137
origin of term, 69, 71
political ideology and, 16–17
as political topic, 173–174
population levels, 139–140, 248
as predators, 114
primary diet of, 137, 140–141
pronghorn antelope and, 184
pronunciation and, 16–17
public opinion of, 15–17
Pueblo Indian rock art, 50
pup survival rates, 197
resiliency of, 15, 103–105, 108, 228–229
rodent control by, 83–84
role of, 20
romanticizing, 113
Ruxton on, 73
as scavengers, 131–132
scorched-earth policy against, 102–103
Seton on, 104
sheep and, 143
sheepmen on, 87
Smith on, 107, 126–128
as social, 35–36
as source of confusion, 15
southward movement of, 4
speed of, 132
sport-hunting, 182–183
stereotype of, 75–79, 84
survival of, 101–102
territory extension for, 145
Twain on, 76–78
vision in, 106
war on, 153

wariness of, 106–108
wolves and, 113–114, 126–130, 133–134, 215
in Yellowstone National Park, 127–129, 139–140, 148
See also Bison Peak Pack; prairie wolf; Rick Creek coyote; urban coyotes; Western coyote
Coyote's Journal, 235–236
The Coyote's Lament (Disney), 151–152, 186
Coyotism, 27, 42
disorder in, 49
power available through, 49–50
West Coast, 235
See also religions
coyotl, 4, 9–10, 58
origin of term, 70–71
See also coyotes
Coyotlinahual (coyote sorcerer), 10
Coyotlinauatl (coyote god), 10
coywolf, 210–211, 224, 227
Crabtree, Bob, 133
Crania Americana (Morton), 222
Crows, 25
Crystal Creek pack, 130
Curwood, James Oliver, 92
Custis, Peter, 60

Darling, Ding, 144
Darwin, Charles, on dogs, 92
Davis, Mike, 199
on urban coyotes, 200–201
Dawkins, Richard, 39
DDT, 156–157
application of, 158
ban on, 159
WWII and, 158
See also poison
De Smet, Pierre-Jean, 49
Demeter, 27–28
Denisovans, 31, 228
Denver Eradication Methods Laboratory. *See* Eradication Methods Laboratory

Denver Wildlife Research Laboratory, 145
Department of Interior, 144
Descartes, René, 91
The Dharma Bums (Kerouac), 235
diseases, of Old World origin, 218
Disney, Walt, 151–152, 179, 186–187
Division of Economic Ornithology and Mammalogy, 95
Division of Predatory Animal and Rodent Control, 108
Division of Wildlife Services, 95, 133, 157, 169
  agribusiness subsidies, 172–173
  continued killing by, 172–173, 247–248
  Fox on, 181–182
  predator control through, 170–171
Dixon, Maynard, 65
Dobie, J. Frank, 9, 19, 141, 179
dogs
  Darwin on, 92
  social lives of, 93
Dorn, Ed, 236
Dorsey, George, 234–235
Druid pack, 130

Earth Day, 163–164
ecological niche
  Grinnell, J., and, 119
  introduction of, 119
  See also niche
ecological philosophy of living, 154
Ecological Society of America, 117, 122
ecology, global age of, 155, 159, 162
The Ecology of Fear (Davis), 199–200
Ecology of the Coyote in the Yellowstone (Murie, A.), 136–138
Edge, Rosalie, 144
"The Effects of Control on Coyote Populations" (Connolly), 148
Endangered Species Act (1973), 167, 171, 186, 216, 221–222
  hatred for, 168
  specifications of, 168–169

Endangered Species Conservation Act (1969), 163
Endangered Species Preservation Act (1966), 162–163
Engelmann, George, 70
Environmental Protection Agency (EPA), 163–164
environmentalism, 161–162
epigenetics, 149
Eradication Methods Laboratory, 98, 102, 124, 172
ethnic cleansing, 115
evolution
  bovine, 6
  of brains, 40
  of *Canidae*, 30–31
  of canids, 29–31, 211, 221
  colonizing mechanisms and, 108, 148
  competition and, 34
  of coyote behavioral traits, 128
  of coyotes, 3, 26, 34–35, 105–107, 132–134, 148
  of gray wolves, 225
  of horses, 6
  of human behavioral traits, 39
  of humans, 29, 31
  morality and, 93–94
  of predators, 34
  sociality and, 40
evolutionary psychology, 39–40
extinction, 122
  accidental, 162

federal policy, science and, 116
federal wolf hunters, 113
feral dogs, 191
First Cause, 37
Fish and Wildlife Service, 95, 144, 221
  humane coyote getter, 146
  on red wolf extinction, 222–223
  taxonomy of wolves, 216
  See also Bureau of Biological Survey
fission-fusion, 35–36, 39–40, 59, 148
  in coyotes, 104–106, 108

in humans, 104–105
population expansion and, 101
Say on, 59
folk tradition, 6
Fonseca, Harry, 244
Food Habits of the Coyote in Jackson Hole, Wyoming (Murie, O.), 136
Ford, Gerald, 169–170
Forest and Stream (magazine), 118–119
Forest Service, 95–96
fossils
of *C. edwardii*, 32
at La Brea Tarpits, 32, 34
Wang's interpretation of, 32
Fox, Camilla, 179, 185, 189
on Division of Wildlife Services, 181–182
education of, 180–181
graduate thesis of, 181
on nonlethal coyote control, 181
Frazier, Ian, 242
Fremont, John C., 62
fur trade
coyote pelts and, 85
predator pelts in, 86

Gabrielson, Ira, 144
retirement of, 145
Galen, James, 100
Gehrt, Stan, 193
funding of, 197–198
on urban coyote ecology, 2, 189, 194–195, 202
General Land Office, 89
genetic purity, 226–227
genocide, 115
Gese, Eric, 172–174
on nonlethal control, 175–177
Ginsberg, Allen, 236
Giving Birth to Thunder, Sleeping with His Daughter (Lopez), 235
Glacier National Park, 119, 125, 148
coyote control in, 100–101
role of, 99–101

God's Dog: Conversations with Coyote (Kitchell), 245–246
golden jackal, 29–30, 143
See also jackals
Goldman, E. A., 118
accomplishments of, 123
on balance of nature, 124
on coyotes, 109, 125, 134, 137
death of, 145
on predators, 109, 123–124
wolves and, 123, 216
A Good Journey (Ortiz), 237
Gotham Coyote Project (Nagy and Weckel), 12
Grand Canyon National Park, 148
Grant, Madison, 223, 227
Gray Wolf Restoration team, 26–27, 130–131
gray wolves, 15, 187
bold behavior of, 106–107
coyotes and, 65–66, 125–126, 214, 217–218
evolution of, 225
migration of, 33
origin of, 216
population reduction of, 101
return of, 128
subspecies of, 216
tracks, 127
Young, S., on, 216
See also wolves
Grazing Service, 95
Great Dog War, 10, 191–192, 197
Great Plains. *See* American Great Plains
Great Smoky Mountains National Park, 224
Greek gods, 27–28
Greeley, Horace, 78
Green Mars (Robinson), 245
Gregg, Josiah, 69
on North American jackal, 71
plains journeys of, 70
on prairie wolf, 71, 85
Grey, Zane, 120

Grinnell, George Bird, 118
  on coyote slaughter numbers, 86
Grinnell, Joseph, 179
  Bureau of Biological Survey and, 123–125
  ecological niche and, 119
  on predators, 121–122, 138
  on public lands, 120
  guild competitor, 194, 198

habituation, 195
Hal (urban coyote), 12
Hall, Jon, 191
Hare, Brian, 93
Harrison, Benjamin, appropriations bill of, 90
Hernandez, Francisco, on coyotes, 57–58
Hetch Hetchy Canyon, 121
Historia Antigua de Mejico (Clavijero), 58
homestead acts, 89
Homo, 26, 31
  as hybrid, 227–228
Hornaday, William T., 108
Huehuecoyotl (Aztec deity), 10, 18, 25, 40
  Coyote Man rock art, 24
  values of, 51
humane coyote getter, 146, 169, 171, 174–175
  Fox on, 181
humans
  behavioral trait evolution in, 39
  Coyote as avatar for, 37, 237, 248
  coyotes and, 2–3, 6, 9, 14–15, 26–27, 191–194, 204–205, 209
  evolution of, 29, 31
  fission-fusion in, 104–105
  Kitchell on, 245
  as observers of nature, 28–29
  as outside nature, 193
  predators and, 76, 106
  as social, 35–36
  wild canines and, 92
  Wilson on, 38–39

von Humboldt, Alexander, 64, 75
hybrid swarms, 219–220
hybridization, 210–211
  attention drawn to, 227
  in *Canis rufus*, 215–216, 219
  forces behind, 225
  genetic markers for, 214
  in *Homo*, 227–228
  size difference and, 224–225
  source of, 220–221
  timing of, 217
  Wayne on, 218–219
Hyde, Lewis, on coyotes, 28

Indian Tales (de Angulo), 235
Ingersoll, Ernest, 78
Irving, Washington, 75
Islam, 27
Isle Royale National Park, 126, 128

jackals, 57
  American species of, 59
  coyotes and, 29, 63–64
  See also golden jackal
Jeffers, Robinson, 236
Jefferson, Thomas, 89, 91
  advisor to, 60
  encouragement from, 54
  as scientist, 55
Jones, Chuck, 237–239
  gallery of, 243
Judaism, 27
Junction Butte wolf pack, 130–131, 134
  See also wolves
Jung, Carl, 235
  on Coyote, 27
  savior for, 37

Kaibab Plateau deer episode, 120, 136
Karok, 25
Kays, Roland, 228
Kern, Richard, drawings of, 67–68
Kern County, CA, 120, 136
Kerouac, Jack, 235
kinship determination methods, 29–30

Kitchell, Webster
  on Coyote, 245
  on humans, 245
Klamath, 25
Klondike Trail, 5
Knowlton, Fred, 148, 172
Koller, James, 235–236
Kruuk, Hans, 76

La Brea Tarpits, fossil assemblages at, 32, 34
A Lady's Life in the Rocky Mountains (Bird), 78
Lamar Valley, 126–129
  control of, 130–131
  coyote packs in, 132
  Dead Puppy Hill, 133–134
  See also Yellowstone National Park
land privatization, 89
The Lands of the Arid Region of the United States (Powell), 90
Leavitt, Scott, 109, 134–135
Lefty (gray wolf), 113
Leopold, A. Starker, 159, 164
Leopold, Aldo, 118, 153, 161, 165
  conversion of, 154–155
  on coyotes in Mexico, 139
  on post-war America, 154
  on predator control, 125
  on predators, 155
Leopold Report, 159–161
Lewis, Meriwether, 51, 95, 109
  animals observed by, 54–57
  on coyote pelts, 85
  on coyotes, 55–57
  explorations of, 54
  journals of, 191
Life in the Far West (Ruxton), 72–73
Linderman, Frank Bird, 235
Linnaean system, 58, 60
Little Ice Age, American Great Plains and, 219
livestock industry, coyote control and, 142–143
Loki, 29

London, Jack, 92
Long, Stephen, 58
Lopez, Barry, 19, 235
Los Angeles
  coyote bite incidents in, 201–202
  ecology of, 199
  fatal coyote attack in, 200
  urban coyotes in, 11, 192–193, 195, 198–201
Lotka-Volterra equations, 120
Loukotka, Rob, 240
Lummis, Charles, 234–235

Maclure, William, 60
Maltese, Michael, 237–239
Man and Nature (Marsh), 89–90
Marsh, George Perkins, 89–90
Mather, Stephen, 100
Maximilian, of Wied-Neuwied, 64, 68
  on prairie wolf, 65–66
McGovern, George, 165–166
McIntyre, Rick, 126, 130–132
Mead, James, 85
The Meaning of Human Existence (Wilson), 39
Mech, David, 126, 173, 226
Melville, Herman, 75–76
Menominee, 25
Merriam, C. Hart, 95, 122, 134
mesopredator release, 143
Mexican War, 66
Mexico, coyotes in, 4
migration paths, 7–8
Miller, Henry, 235
Mills, Enos, on coyotes, 117, 179
Mississippian Culture, 218
Mitchell, Pablo, 243
Mitchell, Taylor, coyote attack on, 209–210, 227
Mollhausen, Heinrich Balduin, 75
Momaday, N. Scott, 236
morality
  canines and, 93
  emergence of, 93

morality *(continued)*
    evolutionary origins of, 93–94
    predators and, 92–93
Morton, Samuel, 222
mouflon sheep, 143
mountain privatization, 89–90
Muir, John
    on balance in nature, 108
    on coyotes, 108
    on nature, 116
Mule Deer Protection Act, 184
Murie, Adolph, 135–136, 179, 207
    analysis done by, 140–141
    beginning research of, 139
    continued coyote fascination of, 141
    evidence collected by, 137–138
    fieldwork vindication, 142
Murie, Olaus, 135–136, 179, 198
    on Bureau of Biological Survey, 147
    continued coyote fascination of, 141
    evidence collected by, 137
    fieldwork vindication, 142
    on research motivations, 142

Nagy, Chris, 12
Nahuatl, 70–72
Narratives of Two Journeys in the Prairie (Pike), 68
National Association for the Advancement of Science, 90
National Environmental Policy Act, 164
National Forest System, 90
    grazing in, 96
national parks
    establishment, 120–121
    predator refuge in, 94, 96, 121–123, 135, 148
National Park Service, 90
    establishment of, 91, 121
    Wildlife Division of, 136
National Wildlife Refuge, 90–91
National Wildlife Research Center, 202
nature
    of animals, 92–93

balance of, 108
humans as observers of, 28–29
humans as outside, 193
Muir on, 116
1950s attitudes toward, 153
Thoreau on, 115
See also balance of nature
Nature Faker Controversy, 92–93
nature lovers, predators and, 141–142
Navajos, 25, 45
    on coyote power, 49–51
    Long Walk, 50
    ma'ii, 49
    U.S. treaty with, 51
Neanderthals, 31–32, 228
Neolithic Revolution, 27
neuroscience, 40–41
Nez Perce, 37
    "Coyote and the Shadow People," 46–48
Nezahualcoyotl, 10
niche
    balance of nature and, 119
    for midsize predator, 119–120
    vacancies, 119
    See also ecological niche
Nixon, Richard, 163–165
    Endangered Species Act of 1973 and, 167
    February 1972 address to Congress, 165
nongame varmints, 153
North American jackal, 67–68
    Gregg on, 71
North American mammoths, 115
North American Quaternary Canines (Nowak), 215
novelty-seeking genes, 40
Nowak, Ronald
    dissertation of, 215, 221
    on red wolves, 215–216, 221
Nuttall, Thomas, 10
    education of, 60
    exploration routes of, 61–63

Old Crip (coyote), 115
Old Man America
　values of, 51
　See also Coyote
Old Man Coyote. See Coyote;
　Huehuecoyotl; Old Man America
The Old Ways (Snyder), 235
Old Woman Coyote. See Coyote
On Deep History and the Brain (Smail),
　40
On Human Nature (Wilson), 38–39
Ortiz, Simon, 236–237
Otis (urban coyote), 11–12, 199

PARC. See Predatory Animal and
　Rodent Control
Parkman, Francis, 62, 75
The Passing of the Great Race (Grant),
　227
Peale, Titian Ramsay, 60–61
Pike, Albert, 68
Pinchot, Gifford, 95–96, 118
Pinker, Stephen, 92–93
Pleistocene era, 3, 7
Pleistocene Extinctions, 28, 34
poison, 36, 85–88, 116, 120, 134, 136,
　160–161, 174–175, 179, 186,
　197, 215
　arguments surrounding, 142–147
　bait stations, 122, 124–125
　ban on, 16, 165–166, 169
　campaign, 152–153
　Carhart on, 157–158
　Carson on, 156–159
　collars, 171, 176–177, 181
　efficiency of, 98–104
　extermination through, 7, 83
　lawsuit against, 164
　policy on, 109
　presidential decisions on, 169–172
　See also DDT; sodium fluoroacetate;
　　strychnine; thallium sulfate
"Poisons—the Creeping Killer"
　(Carhart), 157
political ideology, coyotes and, 16–17

Powell, John Wesley, 90
prairie wolf, 59–64, 73
　Audubon on, 74
　Gregg on, 71, 85
　Maximilian on, 65–66
　Woodhouse on, 67
　See also Canis latrans
predator control, 159–160, 166–167
　through Bureau of Biological Survey,
　　95
　Division of Wildlife Services and,
　　170–171
　Leopold on, 125
　in Marin County, 181
　in Yellowstone, 100
　See also coyote control
predator observation culture, 130
Predator Research Facility, 172, 226,
　233
　inmates at, 178
predators
　American Society of Mammalogists
　　on, 109
　apex, 88
　balance of nature and, 120
　benefits of, 122
　bounties on, 87–88, 96
　eradication appropriation, 97
　evolution of, 34
　folk position on, 118
　fur trade of, 86
　Goldman on, 109, 123–124
　Grinnell, J., on, 121–122, 138
　hatred of, 87
　humans and, 76, 106
　identification with, 13
　keystone, 101
　Leopold on, 155
　midsize, 119–120
　morality and, 92–93
　national park refuge for, 94, 96,
　　121–123, 135, 148
　nature lovers and, 141–142
　rodent control by, 122
　role of, 117–118, 153, 155

predators *(continued)*
    sociality of, 36
    Storer on, 121–122
Predatory Animal and Rodent Control (PARC), 124, 138, 156, 161
    budget, 147
    pressure on, 142
problem-solving abilities, 18
Progressive Era, 98
Project Coyote, 178–180, 182–185, 189
public domain policies, 89–90
public-lands system, 118
    Grinnell, J., on, 120
Pueblo Gods and Myths (Tyler), 28
Pueblos, 44
    coyote rock art, 50
Puff (gray wolf), 131

rabies, 191
Rabon, David, 224
Ragged Tail (gray wolf), 131
Rags (gray wolf), 113
Reagan, Ronald, 170–171
Red Mars (Robinson), 245
red wolf, 214, 219–220
    captive breeding program, 222
    Fish and Wildlife Service on extinction of, 222–223
    genome of, 217–218
    hybridization with coyotes, 214, 219–222, 224–225
    Nowak on, 215, 221
    range of, 216
    recovery, 216–217
    Wang on, 217
    Wayne on, 214, 217–218, 226
    See also *Canis rufus*; wolves
Red Wolf Recovery Program, 222, 224–225
Redington, Paul, 124
religions, 27, 42
    See also Coyotism
Rick Creek coyote, 114
    See also coyotes
Riley, Seth, 193

Robinson, Kim Stanley, 244–245
Rocky Mountain National Park, 148
Roosevelt, Teddy, 90–91, 118
Roughing It (Twain), 76–77, 117, 140, 151, 238
Russell, Osborne, 49, 139
Ruxton, George Frederick, 72
    on coyote pelts, 85
    on Indians and coyotes, 73
    on mountain men name for coyotes, 73

Sabin, Edwin, 78
Salish, 25, 37
A Sand County Almanac (Leopold), 154–155, 165
Santa Fe National Forest, 94
sarcoptic mange, 205
    among wolves in Yellowstone, 131
    as coyote control in Glacier, 100
    Montana introduction of, 88
Say, Thomas, 61, 65
    on *C. latrans*, 58–60
    on fission-fusion, 59
    journal entries of, 59–60
    trapping techniques of, 198
    Woodhouse and, 66–68
Schlickeisen, Rodger, 179
science, federal policy and, 116
scientists, as public relations failures, 144
selfish gene hypothesis, 39
Seton, Ernest Thompson, 51, 92, 103, 116, 179
    on coyotes, 104
sexual fitness, 39
sheep
    coyotes and, 143
    declining numbers of in America, 173
sheepmen
    on coyote control, 102, 142–143
    on coyotes, 87
    Wildlife Services and, 173, 179
Shelford, Victor, 117

Sierra Club, 116
Silent Spring (Carson, R.), 156–159
Silko, Leslie Marmon, 236
Sitgreaves, Lorenzo, 66
Smail, Daniel Lord, 40
Smith, Doug, 133
  accomplishments of, 126
  on coyotes, 107, 126–128
Snyder, Gary, 235
  on Coyote, 236
Society for the Prevention of Cruelty to Animals, 92, 192
sodium fluoroacetate, 156–157, 170
  approval of, 145
  ban on, 166
  effectiveness of, 146
  overreach of, 161
  See also poison
Sonoran desert, 4
species cleansing, 115
species recognition, 225–226
sport hunting, 97–98
standard of living, 155
sterilization costs, 175–177
stockmen, 113
Storer, Tracy, on predators, 121–122
strychnine, 99, 145–146
  availability of, 88
  ban on, 166
  effectiveness of, 85–86, 98
  production of, 85
  See also poison
synanthropic species, 193

Taylor, Edward, 135
Taylor Grazing Act, 135
Teller, Henry, 99
Tenochtitlan, 9–10
Ten-Year Bill, 109
Teton Game Reserve, 173
Texas land ownership history, 147–148
thallium sulfate
  approval of, 145
  ban on, 166
  effects of, 146

time release action of, 145–146
  See also poison
theory of mind, 91–93
Thoreau, Henry David, on nature, 115
Three-Toes (gray wolf), 113, 121
Tisdale, Arthur, 94–95
"Tito: The Story of the Coyote That Learned How" (Seton), 104
Toelken, Barre, 49
totem animals, 19
Townsend, John Kirk, 63
trappers, 113
Travels (Bartram), 214
Travels in the Interior of North America (Maximilian), 66
Treaty of Guadalupe-Hidalgo, 66
Trickster Makes This World (Hyde), 28
Twain, Mark, 20, 51, 117, 151, 238
  on coyotes, 76–78
Tyler, Hamilton, 28

Udall, Stewart, 159–160, 162–163
Ultimate Cause god, 37
Unaweap (gray wolf), 113
United States Geological Survey, 90
urban coyotes, 13–14, 153, 227
  in American cities, 10–11
  in Arizona, 11
  attitude changes toward, 201–202
  Breck on, 203–204
  car accident percentages, 195
  cat-killer legend, 197–198
  in Chicago, 194–195, 202
  colonization mechanisms for, 191–193, 205
  Davis on, 200–201
  in Denver, 11, 203
  diet of, 197–198
  diseases of, 205
  ecology, 2–3
  Gehrt on, 2, 189, 194–195, 202
  Hal (coyote), 12
  hazing, 190
  in Indian America, 10
  individuality in, 194–195

urban coyotes *(continued)*
  intelligence of, 195
  in Los Angeles, 11, 192–193, 195, 198–201
  in New York City, 11–13, 192
  Otis (coyote), 11–12, 199
  phenomenon of, 8–9
  as population control, 198
  pup survival rates, 197
  as symbols of disorder, 199–200
  territory establishment by, 195–196
  tolerance of, 204
  See also coyote

Van Valkenburgh, Blaire, 214
"The Varmint Question" (Leopold), 154
The Viviparous Quadrupeds of North America (Audubon), 75
The Voice of the Coyote (Dobie), 141

Walker's Mammals of the World (Nowak), 216
Wang, Xiaoming, 30–31, 217
  fossil interpretation by, 32
  on red wolf, 217
Wasco, 25
Wayne, Robert, 214, 226
  comparative genetic studies by, 217
  on hybridization dating, 218–219
  on wolf evolution, 225
Weckel, Mark, 12
Western coyote, 77
Whalen, Philip, 236
Whitehouse, Joseph, 55
Whitey (gray wolf), 113
Wichitas, 25
wild canines, 227
  debates about, 213
  eastward migration of, 219–220
  humans and, 92
  nuanced ideas about, 153
  See also canines
wild creature appreciation, Yale poll on, 15–16

Wilderness Act, 162
Wilderness Society, 142
Wilson, Edward O., on human nature, 38–39
Wilson, Paul, 225–226
"Winyan-shan Upside-Down" (Sioux), 42–44
Wisconsin Ice Age, 28, 48
The Wolf Hunter's Guide (Corbin), 92
wolves
  in Asia, 24
  bounties on, 87–88, 96
  coyotes and, 113–114, 126–130, 133–134, 215
  extirpation of, 6, 36, 88, 101, 125, 127, 162
  Fish and Wildlife taxonomy of, 216
  genetic analysis of, 26–27
  Goldman and, 123, 216
  in 1920s, 114
  observations of, 62
  pursuit of, 114
  sociality of, 36
  Wayne on, 225
  in Yellowstone National Park, 126–130
  See also Canis rufus; Crystal Creek pack; Druid pack; gray wolves; Junction Butte wolf pack; Puff; Ragged Tail; red wolf
Wolves in Relation to Stock, Game, and the National Forest Reserves (Bailey), 96
The Wolves of North America (Young, S., and Goldman), 118
Woodhouse, Samuel Washington
  observations of, 66
  on prairie wolf, 67
  Say and, 66–68
Woodruff, Israel, 66
World War II (WWII)
  advanced industrial societies after, 155
  chemical knowledge following, 116, 146, 156

coyote control during, 145
DDT and, 158

Yanas, 45
Yellowstone Gray Wolf Restoration Project, 126
Yellowstone National Park, 99–100, 114, 125, 207, 213
  coyotes in, 127–130, 139–140, 148
  predator control in, 100
  wolves in, 126–130
  See also Lamar Valley

Yellowstone Wolf Project, 126
Yosemite National Park, 121
Young, Julie, 172–173, 226
  on coyote control, 174–175
  on individual variation, 177–178
Young, Stanley, 19, 118, 124
  as coyote hunter, 99
  on gray wolves, 216
  success of, 99
Yukon River, 5

Zion National Park, 148

Photo: Sara Dant

Dan Flores is the A. B. Hammond Professor Emeritus of Western History at the University of Montana and the author of ten books on various aspects of western US History. Flores lives just outside Santa Fe, New Mexico.